Remote Operating for Amateur Radio

Ham Radio, the Internet and Remote Station Control

Author
Steve Ford, WB8IMY

Production Staff

Jodi Morin, KA1JPA, Layout

Michelle Bloom, WB1ENT, Production Supervisor

Maty Weinberg, KB1EIB, Production Coordinator

Sue Fagan, KB1OKW, Cover Design

David Pingree, N1NAS, Illustrations

Copyright © 2010 by
The American Radio Relay League, Inc.

*Copyright secured under the
Pan-American Convention*

All rights reserved. No part of this work may be reproduced in any form except by written permission of the publisher. All rights of translation are reserved.

Printed in the USA

Quedan reservados todos los derechos

ISBN13: 978-0-87259-092-2

First Edition

Table of Contents

Acknowledgements

Chapter 1: The Case for Internet Remote Control
Remote control is increasing in popularity because of dramatic shifts in our society and in technology. The FCC has weighed in on the issue. So has the ARRL.

Chapter 2: Networks and the Need for Speed
An easy tutorial on networking at the Internet level.

Chapter 3: Bring the Internet Home
Setting up a home network for remote station control.

Chapter 4: Hardware Integration
How to go about connecting your station components to your computer.

Chapter 5: The Audio Challenge
Remote control is about more than passing commands to a radio. You have to transport audio as well.

Chapter 6: The Listening Post
Start simple with a remote-controlled receiving station.

Chapter 7: Building a Complete Remote Station
Suggestions (and diagrams) for complete "transceive" stations, plus a discussion of software options.

Appendix

Index

Foreword

Once upon a time, almost every amateur had the ability to set up a station at his or her home with few, if any, complications. Neighborhoods seemed more tolerant of Amateur Radio antennas, especially before the days when homes were seen as investments rather than simply places to live.

But as our author, Steve Ford, WB8IMY, points out, times have changed. Now a growing number of amateurs are discovering that it has become extremely difficult, if not impossible, to set up home stations. Some home owner associations are banning ham antennas of any sort. Whole communities are attempting to pass ordinances restricting Amateur Radio antennas (the ARRL is opposing these efforts). And to make matters worse, neighborhoods are awash with consumer electronic devices that generate inference and seem overly susceptible to our signals.

Fortunately, the advent of high-speed Internet access has provided a possible solution. Hams who suffer from onerous restrictions can establish remote stations in "friendlier" areas and operate them by Internet remote control. That's what this book is all about.

If you're finding it difficult to enjoy ham radio because of restrictions where you live, *Remote Operating for Amateur Radio* is your guide to freedom. In this book you'll learn how to set up a remote station and operate it just as though you were sitting in front of the radio. You'll also become acquainted with the FCC rules that govern remote operating, as well as the rules that impact your pursuit of contests and awards.

No matter what your living circumstances may be, the full enjoyment of Amateur Radio is still as close as your nearest Internet connection.

David Sumner, K1ZZ
ARRL Executive Vice President
Newington, Connecticut
April 2010

Acknowledgements

One of the early pioneers in Internet remote control of Amateur Radio stations was also my "fact checker" for this book: Bob Arnold, N2JEU. Bob is well versed in the ways of the Internet and networking in general. His assistance was invaluable.

Another pioneer, Stan Schretter, W4MQ, was also a font of helpful knowledge. Stan is the original author of the *Internet Remote Tool Kit* and *Internet Remote Base* software, which are both discussed in this book.

A number of amateurs are entrepreneurs who make equipment that is highly useful for Internet remote control. Several of them contributed images of their products, along with other information, for use in this book: Rob Locher, W7GH; J. Pablo Garcia Jimenez, EA4TX; Fred Glenn, K9SO; Larry Phipps, N8LP; Mikael Styrefors, SM2O and Steve Elliot, K1EL.

About the ARRL

The seed for Amateur Radio was planted in the 1890s, when Guglielmo Marconi began his experiments in wireless telegraphy. Soon he was joined by dozens, then hundreds, of others who were enthusiastic about sending and receiving messages through the air—some with a commercial interest, but others solely out of a love for this new communications medium. The United States government began licensing Amateur Radio operators in 1912.

By 1914, there were thousands of Amateur Radio operators—hams—in the United States. Hiram Percy Maxim, a leading Hartford, Connecticut inventor and industrialist, saw the need for an organization to band together this fledgling group of radio experimenters. In May 1914 he founded the American Radio Relay League (ARRL) to meet that need.

Today ARRL, with approximately 155,000 members, is the largest organization of radio amateurs in the United States. The ARRL is a not-for-profit organization that:
- promotes interest in Amateur Radio communications and experimentation
- represents US radio amateurs in legislative matters, and
- maintains fraternalism and a high standard of conduct among Amateur Radio operators.

At ARRL headquarters in the Hartford suburb of Newington, the staff helps serve the needs of members. ARRL is also International Secretariat for the International Amateur Radio Union, which is made up of similar societies in 150 countries around the world.

ARRL publishes the monthly journal *QST*, as well as newsletters and many publications covering all aspects of Amateur Radio. Its headquarters station, W1AW, transmits bulletins of interest to radio amateurs and Morse code practice sessions. The ARRL also coordinates an extensive field organization, which includes volunteers who provide technical information and other support services for radio amateurs as well as communications for public-service activities. In addition, ARRL represents US amateurs with the Federal Communications Commission and other government agencies in the US and abroad.

Membership in ARRL means much more than receiving *QST* each month. In addition to the services already described, ARRL offers membership services on a personal level, such as the Technical Information Service—where members can get answers by phone, email or the ARRL website, to all their technical and operating questions.

Full ARRL membership (available only to licensed radio amateurs) gives you a voice in how the affairs of the organization are governed. ARRL policy is set by a Board of Directors (one from each of 15 Divisions). Each year, one-third of the ARRL Board of Directors stands for election by the full members they represent. The day-to-day operation of ARRL HQ is managed by an Executive Vice President and his staff.

No matter what aspect of Amateur Radio attracts you, ARRL membership is relevant and important. There would be no Amateur Radio as we know it today were it not for the ARRL. We would be happy to welcome you as a member! (An Amateur Radio license is not required for Associate Membership.) For more information about ARRL and answers to any questions you may have about Amateur Radio, write or call:

ARRL — the national association for Amateur Radio
225 Main Street
Newington CT 06111-1494
Voice: 860-594-0200
   Fax: 860-594-0259
   E-mail: **hq@arrl.org**
   Internet: **www.arrl.org/**

Prospective new amateurs call (toll-free):
800-32-NEW HAM (800-326-3942)
You can also contact us via e-mail at **newham@arrl.org**
or check out *ARRLWeb* at **www.arrl.org/**

The Case for Internet Remote Control

Forty-five years before this book was originally written, Bob Dylan released a song (and record album) titled *The Times They Are a-Changin'*. He was singing about social change back then, but the words have since been used to wax poetic about any situation in which large-scale change is taking place – political, cultural, technological, etc.

Amateurs have been witness to many technological changes over the decades. We've watched the evolution from "hollow state" vacuum tubes to solid state components. We have seen the steady progression of microprocessor technology as it shrank our transceivers while adding more features. We've seen Amateur Radio software evolve from crude *BASIC* logging programs to sophisticated digital signal processing.

Throughout history, however, there has been one constant: the Amateur Radio station. By "station" we might mean the traditional depiction of gleaming stacks of hardware in a cozy room with neatly dressed feed lines snaking away to giant outdoor antennas. Of course, a station can just as easily find its place in an automobile, boat or airplane. For many amateurs, a "station" may also be defined as the handheld transceiver they carry in their pockets.

But the times they are indeed a-changin, and for many reasons.

The amateur population is aging. At the time of this writing, the average ham was edging into his or her sixth decade on the planet; many amateurs are older still. Throughout most of their ham careers, the majority took pleasure in their home-based stations. Now, however, an increasing number find themselves "downsizing" their living arrangements, which can mean relocating to apartments or condominiums, or possibly moving to assisted-living environments. Unless they are lucky enough

Many new home developments are governed by Homeowner Associations. These groups tend to enforce strict rules concerning external structures such as Amateur Radio antennas. In fact, most associations prohibit *all* external antennas with the possible exceptions of small parabolic dishes for satellite TV reception. Association restrictions have become serious problems for residents who wish to enjoy Amateur Radio.

to be able to carry on with smaller stations and limited-space antennas, some of these individuals may discover that their enjoyment of Amateur Radio is at an end.

Because of the recent economic crisis, some younger hams have also found themselves "downsized" through job layoffs. These amateurs may have been homeowners, but now they face life as apartment dwellers. Although they may still own station equipment, they lack the necessary space for antennas.

Those fortunate enough to keep their homes and jobs may be living in developments governed by Homeowner Associations that impose restrictions on outdoor antennas (as in *none permitted whatsoever*). These associations are becoming increasingly common.

Other amateurs may simply live on lots that are too small to allow the installation of full-sized antennas, or in homes that were built in exceptionally poor locations (like the budding VHF enthusiast who realizes that living at the bottom of a deep valley is a serious handicap).

And then there are amateurs who discover that their home stations are useless because of electronic interference from nearby neighbors or power lines. Or perhaps *they* are the ones causing interference, to the point where they've had to curtail most of their Amateur Radio operating in order to keep the peace.

The Remote Solution

If it isn't possible to establish or operate a ham station in your home, the next best alternative is to set it up somewhere else and operate by remote control. This is not a new concept in Amateur Radio. Hams have been assembling remotely controlled stations for decades.

From the earliest days amateurs have experimented with wired remote control over relatively short distances. In the beginning these experiments consisted of stations controlled by elaborate systems using small electric motors to manipulate the transmitters and receivers. In more recent times the favored approach was RF control, primarily on UHF frequencies as required under FCC Rules at that time (the FCC has since expanded RF remote control to the 2-meter band).

RF remote control is still a viable option for some, but it is technically and legally complicated. For the RF link to function properly, you must have transceivers at both ends of the path along with all the necessary hardware to pass the commands (and audio) to and from the primary radio. The FCC requires that the control link transceivers automatically identify themselves at least every 10 minutes, which also requires dedicated circuitry to make this possible (Morse code identifiers at the very least).

Of course, the strongest objection to an RF link is the fact that it is limited in terms of usable distance. When you're restricted to exercising your control on VHF or UHF frequencies, there is a practical limit to how far the link radios can be from each other and still enjoy reliable signals. The greater the distance, the more power and antenna gain is needed at both ends of the RF path. Someone who already lives under severe antenna restrictions isn't going to be able to erect, say, a

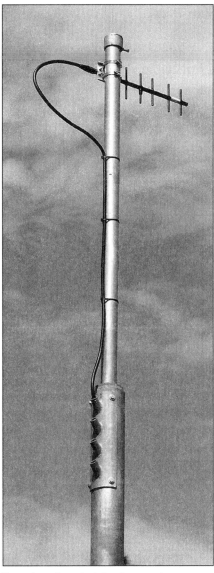

If you're already laboring under severe antenna restrictions, even a small Yagi for RF remote control can become a liability. That's where the Internet has its greatest advantage – no antennas are required at the client location!

long-boom UHF Yagi antenna to communicate with a distant remote station.

Fortunately, the relentless march of technology has provided a much easier solution for remote control – *the Internet*.

Thanks to the Internet, hams now have the ability to remotely operate any station at any distance – all without the complication and limitation of RF linking. The Internet links computers, which are themselves linked to modern transceivers, and it easily transfers large amounts of data between them – including data that carries digitized audio. This has created a boon in remote station control, making it possible for hams to continue enjoying their hobby regardless of where they live. Even local, state and federal communicators are connecting radios to the Internet and controlling them remotely. Among professionals this is known as *Radio over IP*, or RoIP.

One of the first amateur Internet remote stations was created in 2000 by Bob Arnold, N2JEU and Keith Lamonica, W7DXX. Their pioneering work started a revolution in remote station control that continues to this day. In fact, remote operating is growing in popularity, fueled by the fact that Internet access is now almost universal throughout the industrialized world.

And thanks to the popularity of the Internet, personal computers have become ubiquitous. Most homes have at least one laptop or desktop computer in residence. And how many "mobile" computing devices (netbooks, smart phones, etc) are in the market is anyone's guess. Suffice to say that it is a very large number!

How it Works – the Big Picture

The purpose of this book is to offer practical information that will help you assemble your own Internet-controlled Amateur Radio station. Maybe you have a well-equipped home station and you'd like to share your good fortune with some of your friends. Or, perhaps you are a ham living under onerous restrictions and you'd like to establish a remote station so that you can finally get on the air. Either way, a remote-controlled solution is definitely available. All it takes is a certain monetary investment (depending on how much hardware you already own) along with careful planning and a reasonable amount of perseverance.

But before we dive into specific details, let's look at the big picture. In fact, try looking at the picture in **Figure 1.1**.

The station you wish to control is known as the *host*. It's a host because it "hosts" all the gear necessary to transmit and receive RF signals.

If the remote station is the host, that makes the remote operators "guests," right? Well . . . no. In geek-speak, remote users are known as *clients*. Try to commit these terms to memory as they will appear often in this book.

At the risk of being overly simplistic, this is the shopping list for a host station:

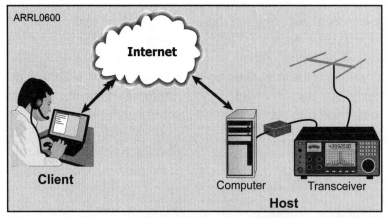

Figure 1.1 – A simplified diagram illustrating Internet remote control of a distant Amateur Radio station.

- A transceiver with a computer interface
- A computer
- Host software
- An Internet connection

A client station typically includes:
- A computer
- A microphone/headset for voice operation and perhaps a CW key
- Client software
- An Internet connection

Host Transceivers and Interfaces

The host transceiver can be almost anything – HF, VHF or whatever. The only requirement is that it has the ability to "talk" to a computer and share it's transmit/receive keying and audio lines.

Fortunately, this capability is found in just about every transceiver manufactured within the last 10 years. Some transceivers offer direct computer interfacing through various ports. Once the transceiver and computer are communicating with each other, the computer can "poll" (query) the rig and obtain information such as the current frequency and band, IF filter selections, RF gain settings and much more. At the same time, the computer can send commands to the transceiver. It can "tell" the radio to switch to a new frequency or mode, increase or decrease output power and so on.

The transceiver control interface usually doesn't transport audio signals. Those are handled separately, typically through a *sound card interface* with cables going between the computer and the transceiver. You will also find devices on the market that combine transceiver control and sound card interfacing in a single package.

The Host Computer

The host computer doesn't have to be anything special or expensive. You're not asking it to store large files or display elaborate graphics. All it needs to do is communicate with the transceiver and the Internet, and run the host software. Technically speaking, the host computer doesn't even need a display monitor. No one at the host station will be looking at it, unless someone needs to access the computer to perform maintenance.

Most amateurs (about 95% at the time of this writing) run computers equipped with various flavors of the *Windows* operating system. Consequently, most host software is written for *Windows*. Be that as it may, you can just as easily use a *Mac OS* or *Linux* computer at the host station. There may even be some security advantages in doing so, which we'll discuss later.

Whichever operating system you choose, a host computer can be remarkably affordable. If you really want to shave costs, you can probably pick up a used desktop PC for a couple of hundred

The Tigertronics SignaLink is typical of the devices known as "sound card interfaces." It allows your computer to switch your transceiver between receive and transmit. It also provides isolation for the audio lines.

dollars that will do the job nicely. On the other hand, prices of new computers in recent years have been plunging like proverbial stones. When this book went to press, for example, it was possible to purchase a brand-new fully loaded Dell computer running *Windows 7* for less than $500.

Internet Access at the Host Location

According to the latest statistics, high-speed *broadband* Internet is available to about 70% of American households. So, if the host computer has access to broadband, you're in luck. High-speed Internet is ideal for remote-control applications because the data can flow between host and client smoothly and quickly.

Broadband access is easily shared, so even if the host station is taking up residence in someone's home, the other members of the household will still be able to enjoy the Internet over the same connection. In all likelihood, they won't even notice client activity. The only potential issue with shared access is that clients may experience sluggish performance if someone in the home is making big demands of the Internet pipeline, such as watching streaming video or downloading a very large file (such as a movie).

If DSL or cable broadband aren't available at the host location, there are other options, which we'll discuss in the next chapter.

What is the FCC's Position on Internet Remote Control?

If you've shopped for FM mobile transceivers, no doubt you've seen models that offer the ability to separate the front panel (or "control head") from the rest of the radio. This is a terrific convenience feature. It allows you to put most of the radio out of sight in the trunk of your car, or perhaps under a seat, while only the lightweight front panel remains in the open. The front panel is much easier to mount than a full-sized transceiver, yet it puts all the control functions and displays at your fingertips while it communicates with the rest of the radio through a slender cable.

The reason I bring this up is because it offers the perfect analogy of how the Federal Communications Commission regards Internet remote control of an Amateur Radio station. To put it simply, the host is the radio in the trunk, the client is the front panel hanging off the dashboard and the Internet is the interconnecting cable.

As far as the FCC is concerned, the Internet is just a *very* long cable. In this situation the FCC's concern is limited to the issue of who controls the host station and how they identify themselves on the air. It doesn't matter where the operator is located; he could be across the street or on the opposite side of the world.

As far as Part 97 of the FCC Rules regarding remote operation is concerned, the Internet is analogous to the cable that links the detachable front panel of a mobile transceiver with the rest of the unit. In this photo of an ICOM IC-2820H transceiver, it is the bundled cable on the left-hand side. As long as the transceiver is operated properly, the FCC could care less about the interconnecting cable. This cable is 10 feet long, but a "virtual" version via the Internet could *10,000* miles long. Functionally and legally, it's all the same.

Manny, NP2KW, maintains a remote control host station in Puerto Rico.

According to FCC Rules, the client is the *control operator*, the person responsible for the proper operation of the host station. The control operator is *not* the person whose call sign is on the license at the host location. In fact, no one needs to be present at the host location *so long as the client is able to maintain control*. The concept of "control" in this sense also requires that the client have the ability to shut down an unattended host transceiver should something go awry. This is a topic we'll discuss in a later chapter.

Since the client is the control operator, the client can only operate the host station within the privileges of his license. For instance, the call sign of the host station may belong to someone licensed as an Amateur Extra, but if the client holds a Technician license he can only operate the host station on frequencies and modes granted to Technicians.

When it comes to call signs, the client is free to use his call sign or the call sign of the host station (if the host station licensee grants permission). In practice, clients usually identify in a specific way to avoid confusion. They will say something like, "This is WB8IMY operating through N6ATQ in Escondido, California."

If we starting talking about Internet remote control that extends across international borders, the legal aspects become more complicated.

■ A foreign amateur can access and operate an US-based Internet remote control station *only* if he is a citizen of a country that has entered into a multilateral operating agreement with the United States, or if he is a citizen of a country that shares a bilateral Reciprocal Operating Agreement with the US. The latest list of participating countries is available on the ARRL Web site at **www.arrl.org/international-regulatory**.

When operating in this fashion, the individual at the client station must identify using "W" and the number of the FCC call letter district in which the host station is based, followed by a slash and his non-US call sign, e.g. W3/G1ABC.

■ An American amateur can access and operate a foreign-based Internet remote control station if the host station is in a country that participates in the European Conference of Postal and Telecommunications Administrations (CEPT) or

International Amateur Radio Permit programs. Again, the list is available at the Web site noted previously.

The American amateur must use the call sign prefix of the country where the host station is located, followed by his call sign. For example, if the host station is in Austria, I would need to identify as OE/WB8IMY. Before you attempt this, however, check the regulations of the host nation carefully.

How Does the ARRL View Internet Remote Control?

When this book was written, the ARRL had not taken a blanket position on Internet remote control per se, except to condone it as a means for hams to enjoy the hobby when they otherwise could not. When it comes to ARRL award programs, the current rules offer guidance.

■**The DX Century Club (DXCC):** Remote-control contacts are valid, so long as the host and client stations are located within the same DXCC entity. For instance, if you live in New York and use a host station in New Mexico to make DX contacts, those contacts are eligible for DXCC credit since both the client and host are within the United States. However, you can't claim DXCC credit for contacts you might make while using a host located in, say, Australia. And credit can only be claimed for the call sign you are using on the air – either the host's or your own.

■**Worked All States (WAS):** Contacts via remote control are valid, but the host and client must not by separated by more than 50 miles.

■**VHF/UHF Century Club (VUCC):** Remote-control contacts are valid, but host and client stations must be within the same grid square.

Those hams who happen to contact remote stations for award credit must understand that the contact counts *only in terms of where the host station is located, not the client operator*. For example, let's say that a client in Ohio is operating a remote station 30 miles away and just over the state line in Indiana. If a ham is hunting his Worked All States award and makes contact with the client, he will earn credit for contact with a station in *Indiana* where the host transceiver is actually located, *not* with Ohio where the client resides. This is why clients should frequently identify the host station location when operating remotely.

As far as contests are concerned, while the rules of each contest vary, the policy on remote receivers is consistent for all ARRL contests. Remote operation is permitted, but all elements (transmitter, receiver, etc) of the host station must be within a 500-meter diameter circle.

There is an "Extreme" category in the CQ Worldwide DX contests that *does* permit remote receivers/operations. Their rules state:

Locations: *The entrant's transmitting sites must be located in a single country, as defined by the applicable licensing authority, and a single zone. Remote receiving sites may be located anywhere.*

See **www.cq-amateur-radio.com/XtremeCQ_WW_Experimenter_June.pdf** for specifics.

2
Networks and the Need for Speed

You don't need to become a computer networking expert to set up an Internet remote-control station. It's helpful, however, to understand as much as possible about how computer networks operate. Not only will the information come in handy during installation, it will be especially helpful if something goes haywire.

Speed and Throughput

When people discuss Internet connections, you'll often hear references to *speed*. What they mean is more technically known as *throughput* — the rate at which data moves from place to place. For the sake of familiarity, however, I'll use the term "speed."

As an Amateur Radio operator you may have also heard the term *baud* or *baud rate* applied to digital communications. Be careful. Baud is a *signaling rate*, the rate at which a digital signal changes states. This is a useful concept in some instances, but not here. That's because baud doesn't necessarily equate to the speed at which data is being transported. When we discuss speed, we'll use the phrase *bits per second*, or *bps* for short. In the computer world we often deal in thousands of bits per second (*kbps*) or millions of bits per second (*Mbps*). You may hear someone refer to the speed of their Internet connection in "meg," meaning millions of bits per second, as in, "I've got a 5-meg line at home."

When we're talking about speed as it applies to the Internet, it is helpful to know that all Internet speeds are not created equal. The fellow who boasts of having a 5-meg line probably doesn't realize that this is simply a designation for the level of service he has purchased from his Internet Service Provider (ISP). Does it mean that he enjoys 5 Mbps data rates to every point on the Internet? Hardly!

When your ISP sells you a subscription to, say, a 5-meg line, you're receiving an assurance that you will enjoy throughput speeds of 5 Mbps, or something reasonably close to that figure, *only between your home and the ISP*. This idea is confusing to some. You'll see them staring impatiently at their monitors, waiting for a Web site to appear, and wondering why their connection is so slow. They don't realize that their connection to the ISP may be perfectly fine; the slowdown is occurring somewhere along the path between their ISP and the target Web site. Alternatively, the distant computer that's supplying the Web site information (known as a *server*) may be running slowly for whatever reason.

To complicate matters, many Internet connections are *asymmetrical* when it comes to speed. This means that the download speed may be quite different from the upload speed. For example, an ISP may sell a 5 Mbps line, but 5 Mbps refers to the *download* speed; the *upload* speed may be only 650 kbps. Most people don't mind the disparity. Since they are mainly interested in downloading material from the Internet, they want the download speed to be as high as possible. For many applications, including remote control of an Amateur Radio station, a slower upload speed isn't a problem. Asymmetrical speeds become issues primarily for applications where you need to move a lot of data in both directions at the same time, such as Internet video conferencing.

The ISP community is slowly moving toward *symmetrical* lines where the up and down speeds are the same, but when this book went to press, most consumer-grade Internet connections were asymmetrical.

Flavors of Internet Access

Even though we touched on this topic in the first chapter, let's re-introduce and expand our discussion of the various "flavors" of Internet access.

Dial Up

Generally speaking, dial-up Internet service, using a conventional telephone line and modem, is the least attractive option for remote station control. Dial-up connections are the slowest and least reliable of the bunch. If you have a clean telephone line and a good-quality, so-called "56k" V.90 modem, the top speed you can achieve is 53,000 bits per second downstream (*from* the Internet), and

33,600 bps upstream (*to* the Internet). In the US, the FCC limits the downstream dial-up speed to 53,000, and the protocol itself limits the upstream speed to 33,600.

This is not to say that it is flatly impossible to use a dial-up connection for remote station control, but it can be extremely frustrating. Dial-up access for either a host or client should be the choice of last resort.

Cable

Many cable-TV companies offer Internet service through the cable system itself, using a *cable modem* installed in each household. This piggybacks two-way digital traffic on special RF channels. In general, cable modem Internet service is an excellent choice. Upstream speeds are usually 128 kbps or better and downstream speeds are almost always at least 1.5 Mbps, often much more depending on how much you are willing to pay. The main disadvantage is that the service can slow down if many of your neighbors are using it at the same time, since the last-mile bandwidth is being shared among many subscribers. It also requires, of course, that you be a customer of your cable TV company, and thus might not be cost-effective if you receive TV service some other way.

Digital Subscriber Line (DSL)

This is a form of high-speed Internet that rides over the existing telephone wiring between your home and the phone company's switching center, or to a device known as a Remote Terminal. Since digital signals can share the line simultaneously with analog signals, you can enjoy DSL service without having to install a second telephone line. With special filters installed on each analog telephone, family members can chat to their heart's content, oblivious to the fact that Internet data is flowing at the same time. (If you don't install the filters, your family will be treated to a loud buzzing noise!) A special DSL modem acts as the interface to the computer.

Like cable Internet, DSL is also an excellent choice for remote station operating. Although its

A typical cable Internet modem.

top speeds are generally less than comparable cable Internet service, the bandwidth is fixed since it isn't being shared with neighbors. So, its actual speed may be higher than cable at many times of the day.

Fiber Optic

Rather than sending information as electrical signals on copper cables, some Internet providers rely on pulses of laser light surging through bundles of glass fiber (hence the name "fiber optic"). You may hear this type of service referred to as *FTTP* – Fiber To The Premises. Verizon Corporation calls their fiber optic system *FiOS* and some are starting to use this term to apply to any form of fiber optic Internet.

If you're lucky enough to live in a neighborhood with this kind of service, give it strong consideration. Fiber optic features very high-speed Internet access, along with telephone and television service, all riding on the same fiber. It is blazingly fast and likely to become the wave of the home Internet future. But the fiber optic rollout has been slow and is confined primarily to major urban areas. It will be years before fiber optic becomes as commonplace as DSL or cable.

Satellite

Several companies now offer Internet service via satellite. For many rural consumers, this is often their only option for broadband access. Satellite broadband commonly takes the form of modems that receive and transmit to satellites in geosynchronous orbit, which means that these spacecraft effectively "hover" over the United States 24 hours a day. The downside of satellite access is the distance. Since you are swapping data with a satellite 22,000 miles away, you are looking at a 44,000 mile path between the host station, the satellite and the satellite's ground station (and, hence, the Internet). This distance is sufficient to cause a slight delay, known in the computer world as *latency*. It isn't likely to have much impact on voice operations, but CW might be problematic.

Satellite communication with typical Ku-band systems is also effected by heavy rain or snow. As satellite TV viewers will tell you, loss of signal during heavy precipitation is a common occurrence. If you live in an area "blessed" with harsh winters, you may also face the problem of ice build-up on the antenna itself. Ice can accumulate to the point where the signal isn't strong enough to reach the detection threshold of the receiver. Fortunately, small dish covers are available for the

Wild Blue is one of several satellite Internet providers.

satellite TV market along with dish heaters and vibrators to keep them clear of ice and snow.

Since Broadband over Power lines (BPL) failed to deliver on its promise of providing Internet service to rural areas, satellite-delivered Internet has become increasingly popular and prices have fallen to affordable levels. To cite an example, when this book was written the WildBlue company was offering satellite Internet with 1 Mbps downloads for $70 per month.

Terrestrial Wireless

"Wireless" has become a catch-all label for any sort of Internet access delivered by ground-based radio systems. Wireless can mean the type of wireless access that you may have as part of your home network, or what you may encounter in a coffee shop, hotel or airport. This sort of wireless access is generally referred to as Wireless Fidelity, or simply *WiFi*.

Wireless can also mean Internet access delivered by cellular telephone companies. So-called *smartphones* such as the popular iPhone communicate with the Internet this way. There are also adaptors available for laptop and desktop computers that will allow them to access the Internet through the cell phone data systems. Data rates are typically on the order of 750 kbps. Cost is an issue with this type of access. Unless you've purchased an unlimited data plan, you are probably paying by the byte, and that can become quite expensive.

Finally, the cutting edge of high-speed wireless is being offered through a technology known as Worldwide Interoperability for Microwave Access, better known as *WiMAX*. WiMAX roll out has been slow to date, and there are several competing systems, so it is not a common means of Internet access, at least so far. Like a cable Internet connection, satellite and terrestrial wireless services are typically shared services. A lot of

This WiMAX installation provides high-speed wireless Internet access.

users or one or two users passing a lot of data through the system are likely to cause slowdowns, or in extreme cases, momentary lost connections.

Networks Big and Small

Sometimes we take it for granted that everyone understands the concept of a *network*. In its fundamental form, a network is any system of interconnected people or things. A human family is a network; so is a baseball team. In Amateur Radio we have groups of stations that meet on the air for a common purpose and we refer to these groups as *nets*, a shorted form of the word "network."

In the computer world, a network is a system of interconnected computers that share information. In "computerspeak", however, we can subdivide the word "network" in several ways. The most common are *LAN* and *WAN*.

LAN is an acronym for Local Area Network. This is a network that exists in a specific area, such as a home or business (**Figure 2.1**). If you've set up a computer network in your house, you have created a LAN. LANs aren't always confined to a single physical space, however. For example, a business may have two office sites, one in New York and the other in Los Angeles. The computers at both locations are interlinked with each other and with both offices. This is still a LAN because it is limited to the company computers.

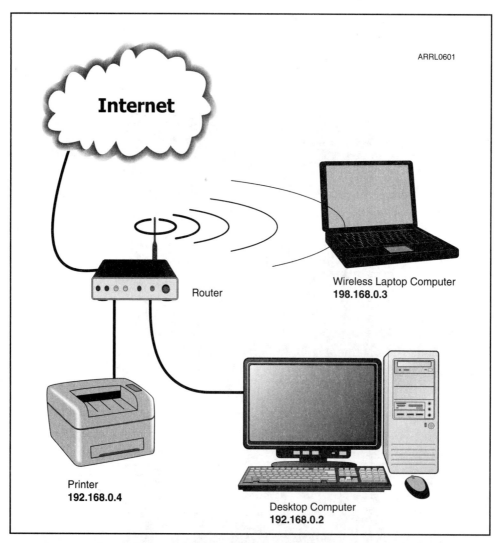

Figure 2.1 – A LAN is a network confined to a specific area, such as your home. In this example, the router acts as the gateway to the Internet. It has assigned unique addresses to each device in the home LAN.

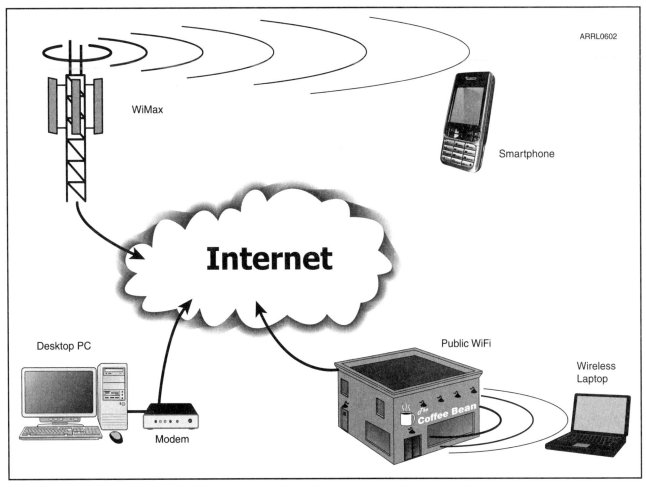

Figure 2.2 – A WAN is a Wide Area Network, linking many types of computers – using many methods of access – over a large area, perhaps over an entire continent.

Once we step outside the confines of our homes or businesses and start interacting with the wider world, we've entered the WAN – Wide Area Network. A WAN is a vast network that is comprised of countless LANs (**Figure 2.2**). It may span cities, states, nations and continents to form a global web of interconnected computers. That sounds an awful lot like the Internet, doesn't it? Well, yes, the Internet is indeed a WAN. It's the mother of all WANs!

What's Your Network Address?

Every computer on a network has an address. So do many devices that have internal microprocessors. These so-called *network-aware devices* may include everything from printers to personal digital music players. To understand how data is moved across a network and shared with all these devices, it helps to get a handle on the subject of addresses.

One way to imagine the Internet is to think of an urban neighborhood with many houses. Think of your own house, in fact. If someone wants to send you a postal letter, they might address it like this:

Bob Johnson
12 Vista Ave
Omaha, NE 68046

This is known as a *hierarchical address*. Disregarding the ZIP code for the moment, a letter processor might look at the bottom line and say, "I'll toss this one into the bin that is going to Nebraska."

Once the letter arrives at a sorting facility in Nebraska, a person (or more likely a robot these days) will say, "This one goes to Omaha."

Finally, a sorter in Omaha gives the letter to a postal delivery person whose route includes Vista Avenue.

Internet Protocol addresses, better known as *IP addresses*, work in much the same way. An IP address is a set of numbers used to locate and identify a device on a network. That device can be a computer, a server, a router or even an Amateur Radio transceiver (assuming it is equipped with a network connection). The addresses are unique within the network. That is, two devices can never have the same IP address. If that happens, the data packets won't know where to go and you'll have a nasty condition called an *IP conflict*.

The IP Address Structure

All IP addresses are made up of four parts (quadrants) separated by dots, like this:

XXX.XXX.XXX.XXX

…where each XXX can be any number between 0 and 255. If you know binary, you will understand that each of these numbers are stored in 8 bits (binary digits), and the number of possibilities you can have is 2 raised to the power of 8, which is 256 (0-255).

Examples of IP addresses are:
192.168.66.5 or 127.0.0.1

The second example above is the default IP address assign to any standalone machine. So, if your machine is not connected to any network, its address is 127.0.0.1. This is also called the *localhost* address.

The Two Parts of an IP Address

An IP address consists of two parts: the network part and the machine part. Let's go back to the analogy of your home postal address. It is made up of the country part, then the city part, then the street part. All people living in your locality will have the same country and city parts in their addresses. Only the house number and street parts will be different.

For IP addressing, all machines on the same network will have the same left-hand (network) quadrants. The final quadrant on the right-hand side is specific to the individual devices. For example, right now I am writing this chapter on a computer within a LAN at ARRL Headquarters. My computer's IP address is 10.15.30.**5** and the guy down the hallway is using a computer that has an address of 10.15.30.**6**. The laser printer in another office is at 10.15.30.**7**. Notice how all three left-hand quadrants are the same? Only the far

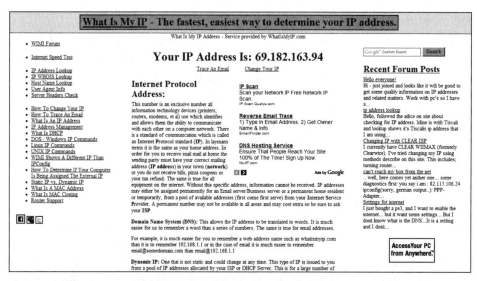

Web sites such as WhatIsMyIP (**www.whatismyip.com**) will instantly reveal your current IP address.

right-hand quadrants are different (shown in bold for emphasis).

Thanks to IP addressing, the data flows through our offices from one device to another, each byte going to its proper destination – just like a postal letter.

Dwindling IP Addresses

An IP address carries 32 bits (8 for each quadrant). This can give up to around 4.3 billion addresses. Unfortunately, many of these are wasted. During the early days of the Internet, big companies bought large chunks of IP addresses and till now can never use all of them. The current version of IP addresses in use is version 4, called IPv4. Since it is predicted that there will eventually come a time when IP addresses will be in short supply, a new version has been developed. IPv5 has been only for research purposes. The next version is version 6, IPv6. This takes 128 bits to store an IP address, so you are sure to get enough addresses for the next few centuries!

How IP Addresses are Assigned

So as not to have any duplication or inconsistencies in the allocation of IP addresses, there is an independent organization that takes charge of IP address allocation. It is called the ICANN (International Company for the Assignment of Names and Numbers). Before the creation of the ICANN in the '90s, there was the InterNIC doing that work.

But as a user, all of this is invisible to you. You don't need to contact ICANN to get an IP address for your home network. Instead, your Internet Service Provider (ISP) does that for you. Chances are, you aren't even aware of your IP address. If you'd like to see it, go to a Web site that displays IP addresses such as **www.whatismyip.com**.

Alternatively, if you are using the *Windows* operating system, try this:
1. Click **START**
2. Select **ALL PROGRAMS**
3. Click **ACCESSORIES** and then **COMMAND PROMPT**
4. When the window pops up, type **IPCONFIG** and press **ENTER**

What you'll see next is the IP address of the computer.

If you're a user of Apple's MAC *OS X* operating system try this:
1. Click on the Apple logo in the top left corner of your screen
2. Click on **System Preferences**
3. Click on **Network** under "**Internet & Networking**"

The Network setup box will appear. Then you can select either the Ethernet port or the Airport (the Mac's wireless connection) to see what your IP address is. A little green ball type icon should appear next to the connection your computer is using.

But if IP address are at the heart of all network communication, why aren't we more easily aware of our individual addresses? The answer is that IP addresses aren't very "user friendly" because they are difficult to remember. The easier solution is to translate these complex numerical addresses into *names*.

Translating Names to IP Addresses

If you have a Web site, it has to have a *domain name,* which, simply said, is what you type to access its main page, e.g. **arrl.org** for the ARRL Web page. Just like IP addresses, each of these domain names have to be unique. You cannot have two sites with the same name and address. The ICANN takes care to ensure that all names and IP addresses are unique.

Whenever you type a Web address into your Web browser, it is quickly converted to its true IP address. How? It is matched to a gigantic catalog of names and addresses stored in Domain Name Servers (DNS) throughout the world. This is accomplished quickly and transparently, at least as far as you are concerned.

For instance, when you type in **www.joe-remote-station.com**, the DNS matches those words to the IP address 192.168.58.5 and, *voila!,* a *virtual connection* is established between your computer and the one at Joe's remote station.

Essentially, the DNS is nothing more than a massive telephone book your computer uses to look up names and determine the required IP numbers. Other than making sure the name points to the correct IP address when a domain name is first registered, it's all pretty much transparent to most users.

You might reasonably ask, "But what happens if the DNS my Internet Service Provider uses goes off line?"

This happens very rarely, but it *does* happen. If the DNS goes down, you're out of luck until it comes back online. Without a DNS, the only way you can make an Internet connection is by entering the numerical IP address into your Web browser. This allows you to automatically bypass the DNS.

It is interesting to note that when a new site comes on line with a new Web address, it can take a while for that information to propagate to all the DNSs throughout the world. This can be time measured in hours or even a day or two. A few years ago, the ARRL Web site had to switch to a new IP address. Until the change propagated through all the DNSs, we received complaints at ARRL Headquarters from members who couldn't access the site, even though their friends could.

Dynamic and Static Addresses

Now that you have a reasonably firm grasp of the concept of Internet addressing, I have one more curve ball to toss your way — and this one is very important when it comes to setting up a host station.

Just like a Web site, your host station needs to have a consistent IP address, one that everyone can remember; one that never changes. The problem is that many IP addresses are *dynamic*, which is another way of saying that they are subject to change without notice. The software technology at the core of this issue is known as *DHCP* (dynamic host configuration protocol). Remember this term because it will appear again later when we discuss home networking.

Internet Service Providers use DHCP because it solves the dilemma of having too many users and not enough IP addresses. In a nutshell, when a DHCP-configured computer or other network device connects to a network, it immediately sends a query that asks, in so many words, "What's my address?" The DHCP server manages a pool of IP addresses and it responds by assigning an IP address, a *lease* (the length of time the assignment is valid), and other IP configuration parameters. Once the address is established, the computer or device is ready to communicate.

In contrast, a *static* address is just what the term implies. It is an address that the DHCP server assigns and maintains permanently; it never changes.

For the great majority of Internet users, the issue of static and dynamic IP addresses is irrelevant. As long as they can access the Internet, they could care less about addressing. Most are not even aware than IP addresses exist.

But if you are setting up a remote station, you care very much! If the IP address of your host station changes daily, your clients will never be able to connect. Imagine trying to deliver a package to a house whose address is different every day and you can appreciate the problem. There are two solutions to this conundrum:

1. Buy a static address. Most Internet Service Providers will be happy to establish a static IP address for you — for an additional monthly charge, of course. Rates vary, but you can expect a static address to tack between $20 and $30 onto your monthly bill. Some ISPs may require you to pay for "Business class" service to get a static IP address. Sometimes a side benefit of paying for the Business class service is that the upload speed may be faster than a normal "home" type connection.

2. Use an address translation service such as Dynip (**www.dynip.com**). For an annual fee (typically about $30), these services track your ever-changing dynamic IP address and automatically route traffic to you. The way it works is rather simple. The service assigns a domain name to you such as W1ABC.Dynip.com. At the same time, you install a small program in your computer. This program constantly "looks" at your IP address. Should your address suddenly change, the program instantly passes this information back to the translation service. If someone attempts to connect to your computer at W1ABC.dynip.com, the service will automatically re-route the connection to your current IP address.

Bringing the Internet Home

We spent much of Chapter 2 talking about how the Internet works in a general way, especially over great distances. Now we're going to zoom in for a closer look at the details of how we connect at home – wherever "home" may be. Our connections can be simple or complicated, but the techniques are the same for clients and hosts.

When it comes down to the brass tacks of setting up your remote-control system, this is one of the most important chapters. By studying this section closely, you may avoid some of the most common headaches as well as some of the risks. (Yes, I said "risks." You'll see what I mean shortly.)

Any "Port" in a Data Storm

Computers communicate to each other over the Internet through the use of *software ports*, better known as simply *ports*. Without getting over our heads in detail, suffice to say that ports are pathways into and out of a computer. There are also hardware ports such as USB, serial (COM), parallel and so on. Those ports are easier to understand because they exist in the physical world, as anyone who has plugged in a USB device will tell you.

But you need to understand how software ports work because they are the means by which you'll control your remote station, or allow others to control it. When data is blocked or otherwise bottlenecked at a port, everything grinds to a halt. Port problems are among the most common, and most vexing, when it comes to establishing a remote station.

For our purposes we'll discuss two software port groups: *TCP* and *UDP*.

TCP stands for Transmission Control Protocol. It sounds complicated, but the concept is simple. Using TCP, the two computers connect directly to each other and remain connected for the duration of the communication.

Look at the example in **Figure 3.1**. During TCP communication, data is formatted into *packets* that are addressed to the destination computer and sent on their way. The safe arrival of each packet is acknowledged by the destination computer before the next packet is sent. If a packet arrives with errors, the destination computer will request a retransmission.

Right about now you may be experiencing a tingle of *déjà vu*, as if you've read a similar description before. Have you ever tried amateur packet radio, or perhaps read a book about it? If so, TCP communication will seem familiar because it works in much the same way. So do some other forms of Amateur Radio digital communication such as PACTOR. If you hear PACTOR communication on the HF bands, you may notice its *chirp-chirp-chirp* cadence. Each chirp is a packet of data and the "answering" chirp is the receiving station either acknowledging its safe arrival, or requesting a retransmission.

TCP's handshaking method of communicating data is terrific for ensuring error-free delivery, but it is a labor-intensive way of getting the job done.

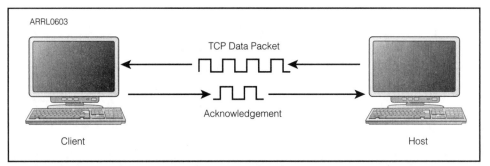

Figure 3.1 – When two computers communicate using TCP, each data packet must be acknowledged by the receiving computer. Either the packet is received correctly, or it is received with errors and must be sent again. TCP guarantees error-free data transfers in this fashion, but efficiency can suffer.

The computers must carefully monitor the communication, check the data for errors or missing packets and respond with the appropriate signals. But when data must arrive 100% error free, nothing else will do. This is the case, for instance, when critical information is being exchanged between the host station and the client operator. If the client commands the transceiver to switch to 14.260 MHz SSB at an RF output of 100 W, you want to be sure the host computer receives the commands correctly. Otherwise, the client could be in for an unpleasant surprise. To paraphrase the old Federal Express slogan, "When it absolutely, positively must get there overnight (and accurately), use TCP!"

UDP stands for User Datagram Protocol. Using this method, the data is still arranged in individual packets, but no attempt is made to confirm their safe arrival and no request for retransmission is sent if a packet arrives corrupted (or not at all). The computer sending the data simply releases its packets into the network with the hope that they will get to the right place. What this means is that UDP does not connect directly to the receiving computer as TCP does, but rather sends the data and relies on the devices in between the sending computer and the receiving computer to get the data where it is supposed to go. This method of transmission does not provide any guarantee that the data you send will ever reach its destination.

This sounds absurd at first blush, doesn't it? Why would you simply kick data packets out the door with no idea whether they'll ever reach their destinations? What would be the advantage of using such a technique?

The advantage of UDP is speed and efficiency because it requires very little computer overhead at either end of the path. Neither computer has to be concerned about whether its precious packets arrived intact. The sending machine blasts them through its ports and into the Internet slipstream. If they show up at the receiving computer, great. But if not, the receiving computer will just wait patiently. Another packet is bound to show up soon. See **Figure 3.2**.

If you're at all familiar with the Automatic Packet/Position Reporting System (APRS) as it is used in Amateur Radio, you may be interested to know that it relies on a similar technique. APRS works by taking position information from mobile GPS receivers, parsing it into packets and then sending it out over the air. Receiving stations decode these packets and display station locations with icons on computer-generated maps. When a receiving station decodes a packet containing new location data, it re-positions the icon on the screen. These bits of information are sent as *unconnected*

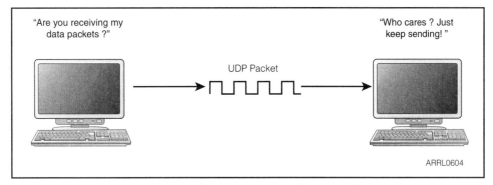

Figure 3.2 – You might think of UDP as a rather "freewheeling" method of data communication. Packets are sent to their destinations and they may or may not arrive intact. It doesn't matter if they show up with errors, or even if they don't show up at all. There is always another packet coming directly on the heels of the one before. As long as you can live with a certain degree of error, UDP can be extremely efficient.

packets in a manner very much like UDP. That is, if a packet arrives with corrupted data, the receiving station doesn't ask for a repeat. Instead, it merely waits for the next packet. This greatly enhances the efficiency of the APRS network by reducing the amount of traffic on the air. Since all APRS stations are on the same frequency, fewer transmissions equal better sharing.

UDP finds its greatest use with applications that stream audio or video content over the Internet. If you watch streaming videos or listen to so-called Internet "radio stations," chances are they're coming to you via UDP. If you use Internet voice applications such as *Skype*, or similar software, they are often employing UDP as well. The secret of UDP's success is the fact that these streams are created using *codecs*. Short for *compression/decompression*, a codec is a special program that reduces the number of overall number of bytes by compressing the data in various ingenious ways. When you process video or audio with a codec, you effectively reduce the amount of data it takes to render viewable images or listenable audio. Some of the data becomes redundant and is "spread" through many data packets. The bottom line is that not every packet has to reach its destination successfully to result in reasonably clear audio or video. In fact, a sizeable percentage can be lost and you'll never notice the difference.

UDP is important to remote-station setups because, as you'll see later in the book, most hams use UDP-based software to carry audio signals between hosts and clients. Once again, however, we need to remember that UDP does not guarantee perfect delivery, and there are times when we *must* have perfect delivery. That's why our software will use either UDP or TCP depending on which protocol is the best tool for the job.

Port Addressing

As we've discussed, every computer or device on the Internet must have a unique IP (Internet Protocol) address. This IP address is used to distinguish your particular computer from millions of other machines. But an IP address by itself isn't sufficient to set up a communication pathway between two computers. Remember that computers use ports to transfer information. To get information into (or out of) a computer, we have to "talk" to the proper ports.

An easy way to understand ports in this context is to imagine that your IP address is a cable converter box and the ports are the different

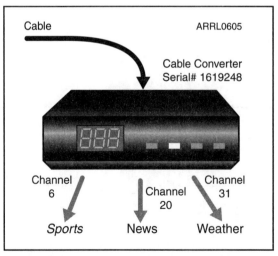

Figure 3.3 – When you think of software ports, consider the ordinary cable TV converter. The cable company sends a stream of many different programs to a single converter, but you receive the ones you want on designated channels. The converter is analogous to a computer and the channels are its ports.

channels on that cable box (**Figure 3.3**). The cable company knows how to send cable programming to your cable box based upon a unique serial number associated with that box (IP Address), and then you receive the individual shows on different channels (ports).

Software ports work the same way. You have an IP address and many ports associated with that IP address. When I say many, I mean *many*. For example, you can have a total of 65,535 TCP ports and another 65,535 UDP ports. When a program on your computer sends or receives data over the Internet it sends that data to an IP address and a specific port on the remote computer, and it receives the data on a (usually) random port on its own computer. If it uses the TCP protocol to send and receive the data, it will connect to a TCP port. If it uses the UDP protocol to send and receive data, it will use a UDP port. Note that once an application connects itself to a particular port, that port can not be used by any other application. It is first come, first served.

This all probably still feels confusing to you, and that is understandable because this is a complicated concept to grasp. Let me offer an example of how this works in real life. Imagine a host station set up in your friend's home 50 miles away. His host setup has a computer that is connected to the Internet 24 hours a day. In order for the host computer to accept connections from a client computer, it must "bind" its host software to a "local port" on the host computer. It will then use this port to listen for and accept connections from client computers. Let's say it is TCP port 5000. The host software connects to port 5000 and waits patiently for connections from clients.

If you sit down before your client PC and decide to get on the air via the host station, the process would work in reverse. Your client software would access the Internet and would attempt to connect to port 5000 on the host computer. Assuming all goes well, your computers connect and data begins flowing.

At this point you must be wondering why you should be slogging through what may seem like such an esoteric discussion. Rest assured that your patience will be rewarded as you plow through the rest of this chapter!

Firewalls, Routers and Security

In the early days of the Internet, computers could usually exchange information freely, simply by knowing each other's host name (or Internet address). No one worried very much about security. But as the Internet expanded and found its way into almost every home, the *cybercriminals* weren't far behind.

One type of criminal is the *virus coder*. This individual (or group) creates small "virus" programs that infect targeted computers. Some viruses are mere nuisances, but others are quite dangerous. The worst among these will search your computer for credit card information, or monitor your keyboard activity, watching for account numbers and passwords. Others will silently recruit your computer to become part of a vast network of machines (known as a *bot net*) that is used to conduct criminal activity.

The most common way to become infected is to open an e-mail attachment that contains a virus. Your computer can also become infected when you visit Web sites that deliberately implant viruses or other malicious codes through your Web browser software. I like to think of myself as technically savvy and yet my station computer was infected after I visited a Web site to look up the lyrics of an old song. Suddenly, my Web browser behaved as if it had a mind of its own, displaying a swarm of unwanted "pop up" advertisements – all this despite the fact that I was using anti-virus software. The nasty code somehow managed to sneak past its defenses and burrow into my hard drive. After a week of trying every cure known to man, I ended up completely erasing the hard drive and then re-installing my operating system and software. Needless to say, I was very unhappy.

The other category of cybercriminal is the *hacker*. This individual thrives on gaining illegal access to computer systems directly. Some hackers simply enjoy the challenge of cracking through a computer's security defenses. Others, however, have robbery in mind. These are the criminals who make news headlines when they break into a corporate system and steal sensitive information such as credit card data.

Hackers use custom software to continually probe the Internet, searching for entry to any computer systems they may discover. They often do this by sending *pings* to random IP addresses. In the computer world "ping" is short for Packet Internet Grouper. It is a tool used for testing networks. It works by sending "echo request" packets to the target hosts and listening for "echo response" replies.

Think of those nail-biting scenes you've seen in war movies. The submarine is creeping through the dark depths, trying to elude the destroyer on the surface. The submariners stare upward, waiting for the sonar pings that will reveal their position (and send depth charges in their wake). When a hacker blasts out a series of pings, he is like the captain of the destroyer, hoping for a response that will reveal his prey. Eventually, a ping will be answered and another unlucky victim will become the target of a hacker.

As an experiment, the Information Technology manager at ARRL Headquarters once "opened" a dedicated computer to the Internet and allowed it to respond to pings. Over a 24-hour period the computer was pinged nearly one thousand times!

This is not to say that every ping represents a hacker with malicious intent. Some computer hobbyists simply like to send pings to see what they can discover. They are content to receive a response and never take it further. Even the malicious hackers tend to be highly selective. Generally speaking, they are not interested in invading home computers. They want to break into large systems such as bank networks. After all, that's where the money is!

Once a criminal hacker locates a potentially attractive victim, he goes to work. The first step is to scan the computer's ports to determine if any are open. Depending on what he finds, these may be the pathways he'll use to gain entry.

Defending Yourself

When you open your host station for access via the Internet, you're facing the possibility of infection or invasion. I wish I could say that setting up a host was a totally risk-free endeavor, but that's not the case. The good news is that there are a number of steps you can take to greatly reduce your risk to the point where it is well within the range of what most reasonable people consider to be "acceptable."

■**Install an effective anti-virus program.** The best way to avoid viruses that arrive in e-mail or through Web browsers is by installing anti-virus software (and by not opening e-mail attachments if you don't know what they are!). The more sophisticated anti-virus programs will scan your e-mail and all other areas of your computer system for threats and will remove whatever they find. The programs also update themselves on a regular basis to stay one step ahead of the criminals.

If you purchase a new computer, chances are it will come with a trial version of anti-virus software already installed. When the trial period ends, buy the software if you are happy with it. It's an investment you won't regret.

Some readers might add that using a computer with a *Mac OS* or *Linux* operating system is a protective measure since most viruses target *Windows* PCs. This is indeed true, but there is nothing magical about *Mac OS* or *Linux* that makes them invulnerable to viruses. The people who create viruses and unleash them on the world are seeking the greatest bang for their antisocial buck. Therefore, they want to see their creations infect the largest number of computers possible. Since roughly 95% of the computers in use today are running a version of the *Windows* operating system, that makes *Windows* systems the prime targets. Some criminal code writers have occasionally turned their attentions to Macs. In 2008 a Mac-specific virus was circulating, although it wasn't particularly destructive. I wouldn't be surprised to see a *Linux* virus one of these days, but with the wild variations in *Linux* operating systems, that's a tougher nut for criminals to crack.

■**Activate or install a *firewall*.** A firewall is a software application that blocks unauthorized access to your computer's ports. If you are using a *Windows XP*, *Vista* or *Windows 7* computer, you'll discover that it has a software firewall already incorporated. Advanced security suites such as those offered by McAfee and many others also include software firewalls. Regardless of who is providing the firewall, *turn it on*. It is one of your best defenses.

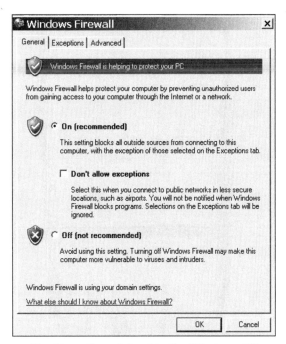

The firewall contained in the Windows XP operating system. Notice that it is on!

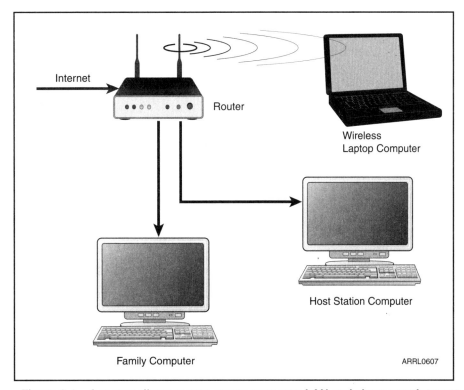

Figure 3.4 – A router allows you to create your own LAN and share your Internet connection with several different users. Here we have a router "dividing" a single Internet connection among three different computers.

This LinkSys router is typical of what you'll find on the market. Selling for less than $100, it offers wired and wireless distribution to your home network.

The only problem with activating a firewall, however, is the fact that it will effectively close most of your computer's ports. That's a deal breaker for remote station control! We'll discuss this issue in more detail later in this chapter.

■**Use a dedicated computer.** By a "dedicated" computer, I mean a computer that is dedicated solely to functioning as the host for your remote-controlled station. Nothing beyond the necessary hosting software exists on its hard drive. No financial records, no personal information, nothing you can't afford to lose.

Used computers are extremely inexpensive these days and, as mentioned elsewhere in this book, you can even acquire brand new computers for less than $500. Unless your host station is working with some hefty applications such as Software Defined Radio, a bargain-basement computer is perfectly adequate. Such a computer still needs a working firewall and anti-virus software, but should the worst-case scenario come to pass, your loss will be minimal.

■**Use a *router*.** A router is, as the name implies, a piece of hardware that "routes" data in a network. These inexpensive devices have become standard equipment in home networks throughout the country. Some cable and DSL modems have routers built-in.

A router allows you to create your own LAN and share your Internet connection with several different users under one roof. For instance, look at **Figure 3.4**. In this example we have a router that is effectively splitting the Internet connection between the family computer and station computer

Most routers arrive with the password set to PASSWORD. Change this to something you're likely to remember and keep a copy on file.

via two lengths of network cable, while at the same time providing wireless access for the family laptop, video game or whatever.

Routers "translate" the main IP address and automatically distribute data by assigning individual IP addresses to each computer or device. This makes it very difficult for a hacker to identify and access the individual machines. Routers also have their own sets of ports, which they will block unless you tell them otherwise (more about this in a moment).

As long as your router is secure, it will function well as your first line of defense. Routers are worthwhile investments even if you don't presently have other Internet users in your household. You never know when a friend or relative will visit with their wireless laptop!

■**Secure your router.** Most routers use User ID and password protection to restrict access to their inner workings. As you'll see later, you will need to use this access to modify your router settings and make them compatible with remote access. The routers often ship with the User ID set to ADMIN and the password set to PASSWORD. You'd be astonished at how many people never change these. Believe me, hackers are counting on the strong probability that you won't change yours, either. Don't give them an easy way into your system! Change the User ID and password, and then keep a record in a safe place.

You may also notice that among your various router settings there is a way to render it unresponsive to pings. If so, use it. By turning off your router's ability to reply to pings, you've just made it invisible to hackers.

Opening Ports

After what you've read so far, the idea of deliberately opening computer ports to the outside world may fill you with dread. There is no way to sugarcoat the fact that continuously open ports will expose your host computer to potential attack. Unfortunately, if you don't open the necessary ports, your remote clients will not be able to connect to the host computer and use the station.

Having said that, if you've followed the previous security recommendations, your odds of becoming a victim are quite low. Hackers are not only looking for lucrative computers to enter, they also tend to look for specific entry ports that your host software probably will not be using. And if

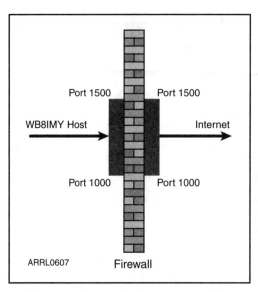

Figure 3.5 – You can "punch holes" in a software firewall to allow traffic to pass through certain ports. In this example, the *WB8IMYHost* program needs to communicate through all ports in a range from 1000 to 1500. *WB8IMYHost* has been added to the firewall's list of authorized programs, so the firewall allows *WB8IMYHost* to use those ports. All other ports remain closed.

you've disabled your computer's (or router's) ability to respond to pings, the hackers are unlikely to find your system anyway. As a personal testimony, I have operated my home PC with a variety of open ports for many years without a single incident. And while no one can guarantee that you'll never find yourself on the receiving end of a hacker invasion, the benefits of remote station control would appear to outweigh the risks.

So, with our fears properly assuaged, how do we go about opening ports?

The answer depends on which ports we're talking about. Your software firewall manages the ports on the computer itself. If you are using a router, it has its own ports. This means that you may have to open two sets of ports.

The firewall software may allow you to specify port numbers, or a range of ports such as 1000 to 1500. The better firewall applications simply require that you specify which programs on your computer should be allowed to open ports upon request. Let's say that I have a bit of host software known as *WB8IMYHost*. By adding *WB8IMYHost* to my firewall's list of authorized programs, the firewall will automatically allow *WB8IMYHost* to open the required ports whenever necessary. See **Figure 3.5**.

Routers work in a similar manner. Depending on the router design, you'll be able to specify the ports, or range of ports, you wish to open. Once again, the more sophisticated routers simplify the process by allowing you to merely add the program to a list of "authorized" software. When the software attempts to open the ports, the router will obey.

But wait a minute…

Didn't we just discuss the fact that routers split the Internet connection, making it available to several users at once by assigning an IP address to each machine? Your remote station clients on the Internet can only "see" and send data to one public IP address. When their data arrives at the ports

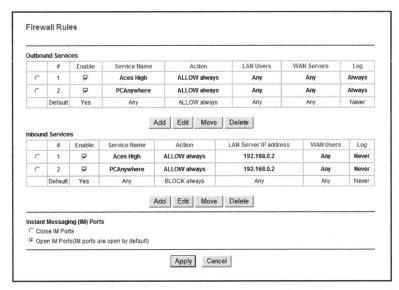

In this example using a Netgear router, notice that it's "Firewall Rules" have been configured to direct inbound traffic for the Aces High or PCAnywhere programs to be sent to a specific computer: the one with the IP address of 192.168.0.2. None of the other computers on this home network will receive that data. This is one form of port forwarding. Each router manufacturer does things a bit differently, but the result is the same.

Routers and Home Network Addresses

It's helpful to think of your home network as the Internet in miniature. In Chapter 2 we discussed the fact that Internet Service Providers assign IP addresses *dynamically* using *DHCP* (dynamic host configuration protocol), which means that your address could change at any time. We also mentioned that it was possible to obtain a *static* address that never changed.

Your home network router works in exactly the same way. A router can assign addresses dynamically using its own DHCP, or you can configure it to assign permanent static addresses to every computer or device on your network.

DHCP works well enough in most instances, but if the router keeps changing the address of your host station computer, this can cause problems. Remember that you're setting up your router to forward data from specific ports to a specific IP address. What if tomorrow the router "decides" to give that address to your spouse's wireless laptop?

The best insurance is to set up a static address for the host station computer so that its address on your network is always the same. Every router has this capability; read your user manual and follow the instructions accordingly.

Bringing the Internet Home   3-11

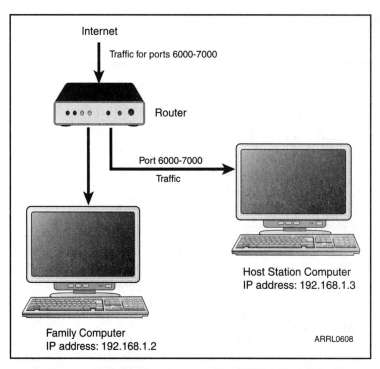

Figure 3.6 – In this example, the router is receiving Internet data traffic intended for all ports in a range from 6000 through 7000. However, this traffic is only intended for the host station computer, not for the family computer. Thanks to port forwarding, you can configure the router to send this port traffic to the host station computer exclusively.

you've opened, how does the router know which computer in the household to send the data to? It doesn't – unless you've configured it to *forward* certain packets of data to a specific machine.

You have to access your router through its Web browser interface to configure its forwarding function correctly. For example, let's say that your host software resides on a computer that has been assigned the following home-network IP address by your router: 192.168.1.3. That host computer is listening for remote clients on TCP ports 6000 through 7000, but it won't receive a single bit of data unless your configure the router to forward all traffic intended for TCP ports 6000 through 7000 to the computer at 192.168.1.3. See **Figure 3.6**. Without a forwarding assignment, the data will sit at the router and end up in the proverbial bit bucket.

If there are hundred different makes and models of routers on the market today, there are a hundred different sets of instructions for setting them up. Most routers have a set of Web-based setup screens, but they all seem to use different commands, and even different terminology, for the various configuration options. Look for options such as "forwarding" or "server setup" in the router manufacturer's documentation. Then, check the firewall requirements in the documentation for the linking system you're setting up, and see if you can put two and two together. Remember that you not only have to forward the router ports to the correct computer, you have to open those ports on the computer's firewall software as well.

Software manufacturers are well aware of what a hassle this can be. Some have implemented so-called *tunneling* protocols that find their way through your firewall and router without any assistance from you. The *Skype* voice-over-Internet application does this with its UDP exchanges. That's why *Skype* is so easy to set up. As far as the user is concerned, he simply installs the software and he is in business. But some applications don't lend themselves well to the tunneling. When in doubt, read the manual.

Hardware Integration

When it comes to wiring a station for Internet remote control, there are no one-size-fits all solutions. The approach you take, and the hardware you need to bring everything together, depends on your ultimate goal. At one end of the spectrum, your ambition may be to set up a simple remote listening post without any kind of transmitting capability. At the opposite end, you may wish to design a complete multiband (and multimode) host station.

In the chapters that follow I'll sketch some rough ideas for various types of remote stations. But for now, let's look at a few *commonalities* you'll encounter. By that I mean elements that would be common to almost every type of station you are likely to set up.

Wiring for Sound

Just about every computer manufactured within the last several years includes sound capability, either through a stand-alone *sound card* or an embedded sound *chipset*. This wasn't always the case, though.

In the early days of home computing — "early" being the '80s and '90s — personal computers didn't normally come equipped with sound devices. They were capable of issuing various beeps through their internal speakers, but that was about all. It wasn't long, however, before the concept of the *multimedia computer* emerged in the marketplace. Consumers were no longer content with beeping machines. They wanted their computers to play their favorite music and videos, and enhance their games with realistic sound effects.

Manufacturers responded with sound cards designed as add-on components, usually plugged into bus slots on PC motherboards. The first sound cards were rather crude, but they evolved rapidly. By the end of the 20th century, sound cards had become sophisticated digital signal processing devices. In fact, the advances in sound card technology ushered in the era of sound-card-based digital modes in Amateur Radio, the most popular of which is PSK31.

By the time we rang in the new century, sound cards had become so common that they had made the jump from frivolous luxury to necessity. Manufacturers responded by integrating sound processing circuitry into the computer motherboards themselves. These are the so-called sound "chipsets." If you're using a consumer-grade computer or laptop made within the last 5 years, chances are it is using the chipset approach to sound recording and reproduction.

In a typical laptop computer, you'll find the sound device making itself evident through jacks labeled MIC and PHONES, or something similar (or perhaps just microphone and headphone symbols). At the risk of stating the obvious, MIC is the audio (microphone) *input* and PHONES is the audio (headphones) *output*.

In a desktop computer you'll find similarly labeled jacks, although the desktop will likely opt to switch the output label from PHONES to SPEAKERS. It may also include a LINE IN jack for higher-level audio input.

Internal plug-in sound cards are still very much

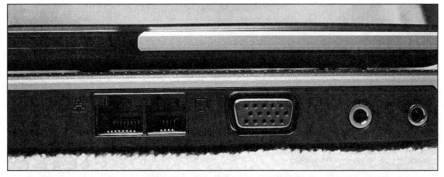

Most laptop computers have two audio jacks: microphone (audio in) and headphone (audio out). In this laptop they are visible on the right-hand side.

A typical computer sound card. Notice the many inputs and outputs.

with us, although you most often see them in more sophisticated desktop computer systems used by gamers and audiophiles. These users demand greater audio fidelity than embedded chipsets normally provide. Sound cards meet this need and, as a result, offer a wider array of input and output options. Even the simplest of sound cards can be complicated these days. They can have as few as two external connections but there may be as many as twelve or more. At the rear of your computer you may find LINE IN, MIC IN, LINE OUT, SPEAKER OUT, PCM OUT, PCM IN, JOYSTICK, FIREWIRE, S/PDIF, REAR CHANNELS or SURROUND jacks, just to name a few. Connections appear not only on the outside, but inside, as well. The main internal connection is to the computer motherboard and this is by an edge connection. There may also be CD audio, telephone, daughterboard and PC speaker

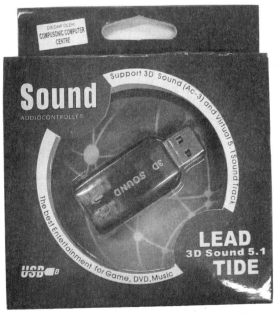

Some external sound devices are as inexpensive and simple as this USB model.

connections inside and on the board as well.

Finally, you may encounter *external* sound devices. Notice that we're not calling them "cards" since they are not circuit boards that reside inside a PC. Instead, these are stand-alone devices that run the gamut from inexpensive modules that plug into USB ports all the way to professional units that are used in recording studios. As you'll see shortly, there are also external devices for Amateur Radio applications that include built-in sound processing.

When it comes to the question of fidelity, the good news is that we're only dealing with voice-quality audio. This means that we can immediately bypass all the expensive high-end sound gear and use whatever is most affordable. Most likely this will be the sound card or chipset that already exists in the computer you've selected as the hub of your host station.

Regardless of the type of the sound device available, its fundamental function is the same. For our purposes the sound device is a tool to convert analog audio from the radio into digital data we can send over the Internet to the client operator. At the same time, it converts the digital audio information from the client into an analog signal for the host radio. See **Figure 4.1**.

Receive Audio

If the host transceiver offers fixed-level audio output at an "accessory" jack, this is the best source for receive audio. You won't have to worry about someone at the host location accidently nudging the transceiver VOLUME control up or down. The audio signal to the sound device remains the same regardless of where the VOLUME control is set.

Most transceivers provide such an output, although they may label it differently. In many instances it is one pin among several in a rear-panel jack. In addition to the receive-audio output, this same jack may carry connections for transmit/receive switching and other functions. As always, when in doubt, consult the manual.

If your radio doesn't have a fixed-level audio output, you'll need to use the external speaker or headphone jack. The problem with using this source is that someone can inadvertently change the output level.

The act of feeding receive audio to your sound device is usually as simple as plugging in a cable. One concern, however, is the possibility of creating ground loops between the computer and radio. A ground loop manifests itself as hum in the audio signal. You can break a ground loop by electrically isolating the radio's audio output from the sound device. This is easily accomplished by adding a small 1:1 transformer (**Figure 4.2**).

Receive audio may not be the only audio you'll choose to share with your clients. A number of transceivers have monitoring functions for voice operating so that the operator can be aware of the quality of the transmitted audio. This audio may be

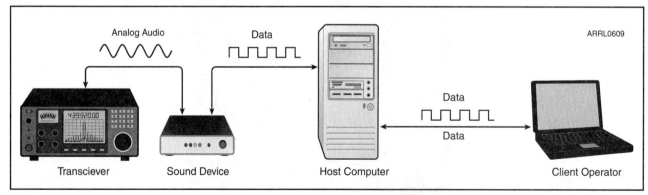

Figure 4.1 – The sound device in this example is acting as a kind of analog/digital translator. It receives analog audio signals from the transceiver and converts them to digital information for the host station computer. The host computer then shoots the digital information over the Internet to the client computer, where it is transformed back to analog audio. When the client operator is speaking into his computer microphone, his computer translates his analog audio to digital information and sends it to the host where the process works in reverse. The sound device can be a sound chipset inside the host computer, an internal sound card, or an external sound interface.

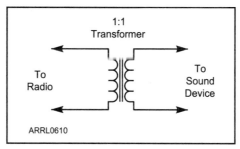

Figure 4.2 – You can use a 1:1 transformer to effectively isolation the audio line between the radio and the computer sound device.

available at the accessory jack, or you may need to tap it from the headphone jack.

Transmit Audio

Just as receive audio goes *to* the sound device, transmit audio comes *from* the same device. When it comes to transmit audio, the most straightforward approach is to treat the sound device output as you would a microphone. That is, wire it directly into the microphone jack of the transceiver. This works well, although, once again, you may need to take steps to isolate the line to minimize ground loops. That 1:1 transformer works in this instance, too.

The alternative is to send the transmit audio to the transceiver's accessory (ACC) jack — if it includes a transmit audio input (not all do). The problem with doing so, however, is that you may bypass the transceiver's speech processing circuitry. Not everyone uses speech processing, so perhaps this isn't particularly important to you.

If you have problems trying to use the radio's rear panel inputs, try the mic jack instead. Some radios have limitations on the use of the rear panel inputs that make their use difficult for remote applications. Among this class of radio is the Kenwood TS-570 and others. While it does have a rear panel ACC input, the firmware within the radio requires the use of the matching PTT input on the ACC jack to enable the input. Some remote host software controls the transmit function through the serial computer jack not the normal PTT input so the ACC input isn't used properly.

Setting Levels

We can't have a discussion about sound devices and audio without delving into the topic of setting audio levels through software. What follows is, by necessity, generic because it will differ from one

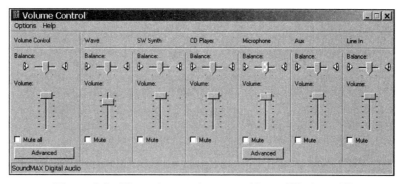

Figure 4.3 – The volume mixer panel from *Windows XP*.

device to another. No two computers are alike, and there are always several different ways to get the same results. In addition, the host software will likely provide its own instructions on setting audio levels. Treat the following only as an overview; the host software instructions may trump whatever you'll read here.

Some sound devices can have more than 35 separate adjustments supported by so-called *virtual control panels*, which is another way of describing the audio controls that appear in separate windows on your computer screen. The sound card may also be controlled automatically through the software programs that you use. There can be many analog and digital signal paths in a typical sound card. Your job is to identify which ones you are connected to and make those connections work. When working in *Windows*, you can single-click the speaker icon in the bottom right corner of your computer screen and see one simple up/down slider with a mute switch. If you double-click that same icon, however, you get a large control panel for playback as shown in **Figure 4.3**.

If you have more than one sound device connected to your computer (an internal sound chipset counts as a "device"), you may have to select the desired device beforehand.

Your control panel may be different. Instead of saying "Volume Control" at the top left corner, it may say "Playback" or it may have "Select" instead of "Mute" boxes at the bottom. Each section has a label indicating the control for a single mono or stereo signal path: Volume Control, Wave, MIDI, CD Player, etc. The word Balance that appears to be part of a section label is not; it is just a label for the left/right slider directly beneath it. This is the playback, output or monitor control panel or, as hams would call it, the transmit control panel. This panel's only function is to adjust what comes out of your sound device. **Figure 4.4** shows a typical "Recording" control panel. The "Recording Control" panel is used to turn on and adjust the input signals from both analog (audio) and digital inputs and to feed those input signals to the software for processing. This functions as a receive control panel. It does not normally control transmit audio.

Depending on your sound device there may be buttons for "Advanced" settings. Here you can find advanced settings for tone control, microphone gain, recording monitoring, etc. Be careful with the

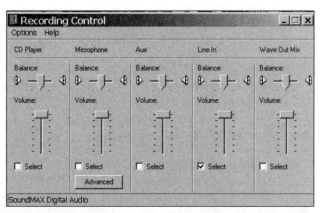

Figure 4.4 – The *Windows* audio recording mixer. Note that the Line Input has been selected and that its gain control is at maximum.

recording monitor settings as they can cause an internal or external feedback loop.

As you look at these controls, note that the sliders are up, the balance is centered; and the inputs and outputs are unmuted. You can be assured that something *won't* work if you have any one of these sliders all the way down or that input muted. Again, your sound device may be different, but if it says "Select" instead of "Mute" you have to check the box instead of unchecking it.

External Sound Devices

The latest trend in sound devices for Amateur Radio has been to create external units that contain their own built-in audio chipsets. In addition to processing audio, these all-in-one devices also provide transmit/receive switching, which we'll discuss next. They are gaining in popularity because they are extremely easy to install and use. These handy units have their own independent audio controls so you don't have to worry about setting the correct levels, or the prospect of those levels changing when you least expect it!

The only potential complication, and it is a small one, is that you may need to tell your host audio software which sound device to use. That's because your computer will "see" both its own sound chipset and the external device. Your host software has to know which one is correct.

We will discuss audio software in detail in the next chapter, but suffice to say that you will probably need to choose which sound device you want the software to use. Many applications make this easy by presenting a drop-down menu that lists all the sound devices your operating system has detected. All you have to do is select the external sound device for both input and output.

Transmit/Receive Switching

Among the many tasks your host computer will handle, one of the most important is transmit/receive switching. If you think of the Internet as a very long microphone cord, the host computer is effectively functioning as the client's push-to-talk (PTT) switch.

Every host software application provides a means to key the transceiver, usually through the computer's serial (COM) or USB ports. The most common way to make the switching connection between the computer and the radio is through a *sound card interface*. These devices appeared in the Amateur Radio market as sound-card digital modes began to flourish and they've become standard equipment in stations throughout the world. Sound card interfaces are available in many different configurations from small and inexpensive to elaborate and pricey.

A typical sound card interface executes its switching function by grounding the transceiver keying line at the rear-panel accessory jack, or the PTT line at the microphone jack. This places the rig into the transmit mode.

The microHAM Interface III is a USB model with its own built-in sound device.

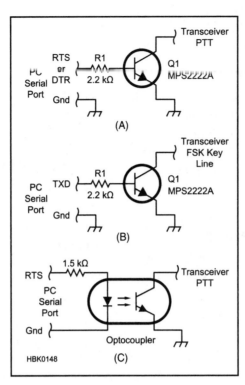

Figure 4.5 — At A, a simple circuit to use the computer COM port to key your transceiver PTT, and at B, a similar circuit for FSK keying. Q1 is a general purpose NPN transistor (MPS2222A, 2N3904 or equiv). At C, an optocoupler can be used to provide more isolation between radio and computer. On a DB9 serial port connector: RTS, pin 7; DTR, pin 4; TxD, pin 3; GND, pin 5.

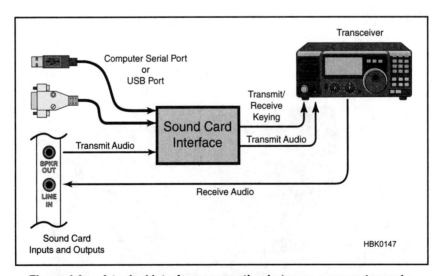

Figure 4.6 — A typical interface connection between a computer and a transceiver. Note that the transmit audio connects to the radio through the interface, and transmit/receive keying is provided by the computer serial port. Newer sound card interfaces are often designed to work with computer USB ports.

This is a straightforward function, so you don't need an expensive sound card interface to get the job done. Of course, the more upscale the sound card interface, the more features it offers. Some models include built-in isolation for the transmit audio line, the receive audio line or both. Others offer front-panel audio gain controls (no more *Windows* "mixer" screens). Check the advertising pages of *QST* and you'll find these products being sold by manufacturers such as MFJ Enterprises, West Mountain Radio, microHAM, Tigertronics and many others.

If you'd prefer to roll your own, you can do that as well. See **Figure 4.5**. This simple keying interface will ground the PTT or other transmit line whenever a logic "high" appears on the designated COM port pin. If your computer doesn't have serial ports (they are fading fast these days!), you can also use a serial-to-USB converter. **Figure 4.6** shows a typical sound card interface connection between a computer and a transceiver.

Speaking of USB, you may discover that some USB devices actually communicate with your computer through *virtual COM ports*. These are created automatically when you attach the USB cable to your PC. Your host software treats it the same as an ordinary COM port, but some applications may need to be "told" which port is the correct one to use. See the sidebar "Understanding Virtual Serial Ports."

It's important to mention that you may be able to use the VOX (voice-operated switching) function

Understanding Virtual Serial Ports

Some USB devices communicate through the use of *virtual serial (COM) ports*. Unlike the hardware serial ports you may find on the back of your PC, virtual ports pass serial data via the USB port and "behave" like hardware ports, at least as far as the software is concerned.

When you install a USB device for the first time, your operating system may attempt to load a *driver*, a piece of software that facilitates communication. If the driver already exists in the operating system (otherwise known as a *native* driver), it will load immediately. If not, you'll be asked to insert the CD that contains the driver.

Once the driver is loaded and running, it will create a virtual serial port automatically. Your operating system will assign a number to the port (COM 5, for instance). In many cases the port assignment is utterly transparent. The software will locate the virtual port on its own and you won't have to lift a finger.

In some instances, however, you'll be asked to select the proper port yourself from within the software. This usually happens during the software setup procedure. If you are presented with a list of COM ports in a drop-down menu, which one is the virtual port connected to your USB device? Unless the software is smart enough to label the COM port list in an informative way, you'll end up having to track down the answer yourself.

This is an easy bit of detective work. In *Windows*, open the **Control Panel** and then double click on the **System** icon. Now click the **Hardware** tab and select **Device Manager**. In **Device Manager** you'll be presented with a list of all the devices and ports that *Windows* can find. Click the + sign to expand the **Ports** list and you'll see the virtual serial port and its assigned COM number. Write this down and then choose the COM port in your program. You're done!

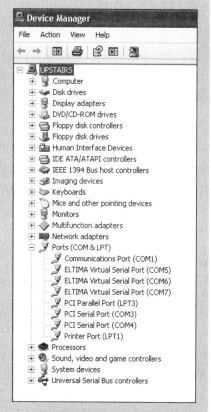

In Device Manager you'll be presented with a list of all the devices and ports that *Windows* can find. Click the + sign to expand the **Ports** list and you'll see the virtual serial port and its assigned COM number.

in your transceiver to automatically switch from transmit to receive when the radio senses the transmit audio from your sound card. This approach completely removes the need for a T/R switching circuit, COM port and so on. The weakness of this technique is that it will cause your radio to transmit when it senses *any* audio from your computer—including miscellaneous beeps, music, etc.

If you're setting up a multimode host station, you'll be happy to know that a number of sound interfaces support not just voice or digital operation, but also CW with separate CW keying lines. Several also support Frequency Shift Keying (FSK) for use with RTTY.

"Talking" to the Radio

Throughout this chapter we've been discussing the means by which we transport audio back and forth between the radio and the computer. One piece of the host puzzle that's still missing is how we go about *controlling* the transceiver from the computer.

By "controlling" I mean more than just transmit/receive switching. This is about establishing a digital conversation between the two devices, allowing the computer to send commands to the radio and the radio to communicate its information to the computer (and, ultimately, back to the client). Such an exchange is possible because modern transceivers contain microprocessor computers of their own. All we're really doing is setting up an interface link so that the host computer can exchange information with the computer in the radio.

You may hear this type of transceiver/computer interfacing referred to as *CAT – Computer Aided Transceiver*. CAT is a term originally established by Yaesu to refer to the control protocol used with its transceivers. ICOM employs a transceiver communication format that they refer to as *CI-V*. CAT, however, has grown to become a generic label for external transceiver control, regardless of the brand in question.

The amount of digital communication that takes place over a CAT link can vary from one radio to another. Some transceivers provide access to channel memories only. This is typical of mobile and handheld rigs. Other transceivers provide a wealth of information to the outside world and allow access to almost every function imaginable.

Transceivers communicate with computers in various ways and there is no established standard. The short list includes…

■**RS-232 (including IF-232 and FIF-232):** This is classic serial data communication with some variations — the same kind you'll find in many computers and other devices. It usually shows up as a 9-pin (DB9) socket on the transceiver's back panel. RS-232 is among the easiest approaches to CAT because all you need is an ordinary serial cable between the radio and the PC. Unfortunately, RS-232 ports are not common features in Amateur Radio transceivers.

■**TTL:** Otherwise known as *Transistor-Transistor Logic*, TTL was the earliest form of transceiver/computer communication and it is still common today. The problem with TTL is that it is incompatible with serial communication. So, to get a computer to talk to a radio with a TTL port you need a "level converter" interface. You'll occasionally see these called "CAT Interfaces." If you purchase these interfaces from the transceiver manufacturers, they will be somewhat expensive. The economical alternative is to buy a third-party converter, or a deluxe sound device that includes a TTL converter.

The connection between the TTL interface and the computer is usually made through a standard serial cable. However, if your computer lacks a serial port, you will need to use a serial-to-USB converter.

■**USB:** Transceiver manufacturers have been rather slow to get aboard the USB bandwagon, but a few have done so in their higher-end radios. If you are lucky enough to own one of these rigs, connecting the radio to the computer is as easy as plugging in a USB cable.

■**Ethernet:** At the time this book went to press, only a couple of Amateur Radio transceivers offered Ethernet ports. One example is the Ten-Tec Omni VII HF transceiver.

Ethernet is the communication standard used in computer networks (the official name is IEEE 802.3). If you have access to the Internet through cable or DSL, the modems connect to your computer through Ethernet ports. If you have a router in your home network, it makes its "hard wire" connections through Ethernet ports as well. Any device with an Ethernet port is *network aware*, which means that it is assigned its own IP address on the network just like any other computer and can be accessed (and controlled) in the same way. As you might imagine, a network-aware transceiver with an Ethernet port is ideal for remote station control. In a later chapter we'll outline what an Ethernet-based host station might look like.

Regardless of how your transceiver makes its connection to the wider world, it depends on software to provide the "human interface." There are many Amateur Radio applications that support transceiver control to varying degrees. When we discuss typical station setups later in the book, I'll provide sources for compatible software.

Controlling the Antenna System

The type of antenna control you require depends, of course, on the type of antenna system you've installed at your host station. If the host station operates on only one band with a single antenna, you don't need to worry about controlling the antenna system at all. This is also true if your host station uses a multiband antenna such as a trap dipole that performs its own "band switching."

On the other hand, if your host station is connected to a proverbial antenna farm, the possibilities become quite a bit more interesting!

For example, you can give your clients access to a rotatable antenna and allow them to aim it in any direction they desire. There are a number of third-party devices that will connect between the antenna rotator controller and the host computer. These devices allow the computer to send commands to the rotator and communicate the antenna position back to the client operator. They typically connect between the rotator control box and the computer through a printer (LPT) port or a serial (COM) port at the computer. Some units install within the rotator control box itself. For example, the Rotor-EZ unit from Idiom Press (**www.idiompress.com/rotor-ez.html**) is an add-on for CD45, Ham-II, Ham-III, Ham IV and TailTwister rotator controllers and it includes an RS-232 option for remote control. The controller sold by EA4TX at **www.ea4tx.com/products/ars-rotators.htm** provides computer control for a wide variety of rotator models. You may even discover that the rotator manufacturer sells a computer interface as an accessory item.

All the interfaces in the world will not do much

The Rotor-EZ unit from Idiom Press shown inside an antenna rotator controller.

The EA4TX rotator controller.

good if your host/client software doesn't support remote rotator control. Not all do, so research the available software carefully. Some applications do not control the rotator directly, but instead launch a separate program to do so. For example, there is N8LP's free *LP-Rotor* software for *Windows* that's designed to be used with the Hy-Gain DCU-1 or RotorEZ interfaces for Hy-Gain rotators. You'll find it at **www.telepostinc.com/LP-Rotor-Html/**.

If you have more than one antenna at your disposal, you'll be pleased to learn that antenna *switching* is straightforward and need not involve the computer at all. There are a number of electronic antenna switches on the market. Some are designed for use indoors at the station console while others are intended for outdoor installation. What many of these have in common is the ability to connect to a transceiver's *band data* output.

Many modern transceivers provide a band data port on the rear panel where certain signal voltages appear according to which band has been selected. This is the "band data." For instance, a pin may have 6 Vdc applied to it when the transceiver is placed on the 17-meter band, but 8 Vdc when it is switched to 15 meters. An electronic antenna switch can use this information to switch from one antenna

N8LP's free *LP-Rotor* software for *Windows* is designed to be used with the Hy-Gain DCU-1 or RotorEZ interfaces for Hy-Gain rotators. You'll find it at www. telepostinc.com/ LP-Rotor-Html/.

to another automatically as a transceiver changes bands.

Note that on some transceivers you won't find a port labeled "band data" per se. Some rigs provide the band information at the port used to connect control lines for linear amplifiers or automatic antenna tuners. Others supply this information on one of the accessory jack pins.

Automatic band switching is the best choice for obvious reasons. In this way, antenna switching is transparent to the client operator and is one less piece of information that has to be communicated over the Internet link. But if the transceiver you are using does not offer a band data port, you can still connect many brands of electronic antenna switches directly to the computer. Here again, however, your host software must have the ability to switch antennas via the COM or LPT ports.

Amplifiers

Is it possible to add a high-power amplifier to a host station transceiver? The answer is … *maybe*. Tuning an amplifier via Internet remote control is a dicey proposition at best, so if you must use an amplifier you need one that is (1) dedicated to a single band and antenna so it never needs tuning, or (2) will switch bands and tune itself automatically.

Modern high-end (read: *expensive*) solid-state amplifiers usually offer automatic band selection. These amps are designed to be as worry free and easy to operate as possible. The instant the radio transmits, the amplifier senses the RF frequency and selects the correct band. This is ideal for remote operating since the client isn't even aware that the amplifier has switched bands.

Not many remote-controlled stations use amplifiers. Cost is a factor, but so is the added complexity and the greater tendency of an amplifier to invoke "Murphy's Law." (If something can fail, it will fail, and always in the worst way.) For example, adding an amplifier increases the RF level in the host environment; computers are notorious for behaving badly when high RF levels are present.

This is not to say that you cannot or should not use RF power amplifiers in your host station. Just be prepared to deal with possible RF-induced glitches. Having a copy of the *ARRL RFI Book* at hand is probably a good idea.

When the Devil Laughs

If you are careful in planning and assembling your remote-control station, you have an excellent chance of enjoying years of trouble-free operation. However, there is an ancient Russian proverb worth remembering: "When we speak of the future, the devil laughs."

Failure is always lurking in the details. We may sincerely believe that we've anticipated and solved every possible problem that could occur, but inevitably we'll be proven wrong. The devil almost always gets his due, one way or the other.

So if failure is unavoidable in a remote-control

If your host station is going to include an RF power amplifier, consider a model that offers automatic band switching.

environment, you need to have mechanisms in place to mitigate the damage. That's why missiles have built-in explosive charges. If the launch officer believes the missile is veering out of control, he hits the ABORT button and sends a signal that triggers the charges. Better to blow a multimillion dollar rocket to bits than watch helplessly as it streaks toward the nearest city!

According to Federal Communications Commission, we need to have an ABORT button for our host station as well. To quote the FCC Rules…

97.213(b) Provisions are incorporated to limit transmission by the station to a period of no more than 3 minutes in the event of malfunction in the control link.

Imagine a scenario in which a client operator is transmitting at the very moment a power surge occurs at the host station. The host computer suddenly freezes and becomes unresponsive to commands. The transceiver continues transmitting and the client is helpless to shut it down. The licensee of the host station is out enjoying dinner with her family and blissfully unaware of what is happening at home. She won't return for hours and there is no one else available to pull the plug!

This is the nightmare that fills many a host station owner with dread. Not only would the station be operating illegally after the 3-minute mark, the transceiver and other components are likely to be damaged after what could be hours of continuous transmission. Taking this to the extreme, it is even conceivable that a device could self-destruct in such a fashion that it ignites a fire.

We can't install missile ABORT switches at the client locations (or explosive charges at host stations!), but we may be able to do the next best thing.

Your first line of defense might be the transceiver itself. A number of transceivers have a feature known as *Auto Power Off*, or something with a similar label. It allows you to specify a transmit time limit. If the operator exceeds the time limit, the radio automatically shuts down, or at least stops transmitting. If your host transceiver offers this feature, set the limit to three minutes.

Some readers may object to what seems like such a short time span, but three minutes is more than adequate for normal conversation. If you don't believe it, start talking and time yourself. You'll discover that you can say a great deal in just three minutes.

Others may point out that the 3-minute rule only applies if the control link malfunctions. This is true,

A DTMF controller such as the Viking RAD-1 (**www.vikingelectronics.com**) will answer an incoming telephone call and allow the caller to deactivate the host transceiver if necessary.

but how does the host computer determine that the control link has indeed malfunctioned? Yes, it is possible to create software that can detect a control link failure, but what if the computer itself fails, as in our nightmare scenario?

The FCC has made it clear that a remotely-controlled station must have a means to shut down the transmitter that is *independent of the control link itself*. This makes perfect sense. It would be foolhardy to design a remote-control station that relied exclusively on its own control link to stop a runaway transceiver. There must be an alternative.

Perhaps the easiest solution is to establish an alternate control line via telephone. This would consist of a DTMF (*TouchTone*) decoder/controller permanently attached to a telephone line at the host station. Ideally, the telephone line would be separate from lines used for voice access or DSL. It would have its own number and would be strictly dedicated to serving as your failsafe link. Should the worst come to pass, the client (or any other trusted individual) could place a call to the dedicated number and punch in a code that would turn off the transceiver.

DTMF controllers are fairly common. For example, there is the Velleman remote-control kit available from Circuit Specialists (**www.circuitspecialists.com/prod.itml/icOid/6205**), or the industrial-grade Viking RAD-1 (**www.vikingelectronics.com**). Ramsey Electronics offers their own DTMF remote-control kit as well (**www.ramseyelectronics.com**). Be careful to check the specifications when shopping for a DTMF controller. Some units are quite sophisticated; they'll "answer" the incoming call, request password entry and even provide a status report. Other units offer only basic features and some require a separate answering machine to "seize" the telephone line in response to a call.

The most direct way to terminate transmission is to have the controller break the power line to the transceiver. See **Figure 4.7**. To do this, however, you'll need a high-current relay in the power supply line; the tiny relays in the controllers won't tolerate the current.

Installing a failsafe system is a bit like buying insurance. With luck, you never need it, but you're awfully glad to have it when you do. It's tempting to play the odds and gamble on the notion that your host station will never run amok. After all, DTMF controllers cost $100 or more. Additional dedicated telephone lines aren't free, either. Why not spare yourself the extra expense?

But running a host station without an alternative

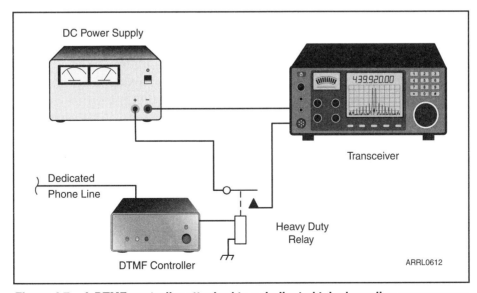

Figure 4.7 – A DTMF controller attached to a dedicated telephone line can serve as your master failsafe. If something goes seriously wrong, you can call the private number, activate the DTMF control and interrupt dc power to the transceiver.

But What About the Client Hardware?

This chapter is devoted entirely to setting up the host station hardware, but aren't we missing something? What about the client?

Client setup is given short shrift because, frankly, there is so little required. The client operator is controlling the host through a laptop or desktop computer that is running the client version of the remote-control software. The only hardware required is a headset/microphone combo such as the one shown here. These headsets may come equipped with separate microphone and headphone plugs that attach to the sound device ports (whatever they may be). Other headsets may connect through a USB jack. In fact, some headset models contain their own sound devices. When you plug them into the PC, they are treated in the same way as sound cards.

Theoretically, just about any headset will do, but it makes sense to invest in a model with the best audio characteristics. Shop carefully and test before buying whenever possible.

If you provide CW capability at your host station, your clients may need an external keying device such as the K1EL Winkeyer (**www.k1el.com**) which uses its own remote-control software to allow CW operation with the client's set of CW paddles. However, a number of host stations tackle the CW issue by having the client simply type text in a software "chat" box, which is then translated to CW at the host and transmitted over the air.

A microphone/headset combo. There are many of these on the market, all designed for computer use.

Remote CW is possible with an external keying device such as the K1EL Winkeyer (www.k1el.com) at the client location. It uses its own remote-control software to allow CW operation with the client's set of CW paddles.

---

control system is asking for trouble. It comes down to a personal choice: spend additional money for a system you may never need, or tempt the devil's sense of humor.

One side benefit to adding a phone line and DTMF remote control is that it could be used to restart the control computer and radio. Computers can be finicky things and the ability to remotely power it off and then back on again can solve a bunch of problems easily. Otherwise a run to the remote site might be needed to reset a frozen PC in the middle of the night.

The Audio Challenge

Shuttling data back and forth between host and client is relatively easy – to a point. For example, the host transceiver sends information to the client only when something has changed. That might be a fluctuation in an S meter reading, or simply a confirmation that a command from the client has been obeyed, such as a change in frequency. The client operator sends even less data, consisting mostly of commands to change transceiver or antenna settings. These communications amount to quick bursts of data, something easily handled over almost any Internet connection.

When we're dealing with audio, however, the story becomes more complicated. Audio data is flowing almost continuously between the host and client. Remember our discussion of UDP a few chapters ago? That's the "unconnected" Internet protocol most often used for audio and video transmissions because it supports a highly efficient, continuous *stream* of information. That is why it is referred to as "streaming" audio or video. In our remote-control scenario, the stream of receive audio (and possibly transmit-monitoring audio) flows from the host to the client. At the same time, and in full duplex, the audio from the client's microphone is flowing to the host. See **Figure 5.1**.

If this audio-handling technique sounds somewhat familiar, you may recognize it by another name: *VoIP* (Voice over Internet Protocol). If your telephone service comes courtesy of your Internet provider, you're making your calls using VoIP. If you hear someone say, "I play an online game with my friends and we talk at the same time," those exchanges took place using VoIP.

VoIP software applications have been around for years and we can easily put them to work for remote station control. Of course, what this means is that you'll probably be running at least two programs simultaneously at both the host and client locations: the station control software and the VoIP application.

But Which Application?

There are many VoIP software applications available. See **Table 5.1**. Most are written for the *Windows* operating system, but there are also versions for *Linux* and *Mac OS*. Some applications are free and open source, while others come at a price and are closed source. So which one is best?

That question is difficult to answer because it depends on their individual strengths and

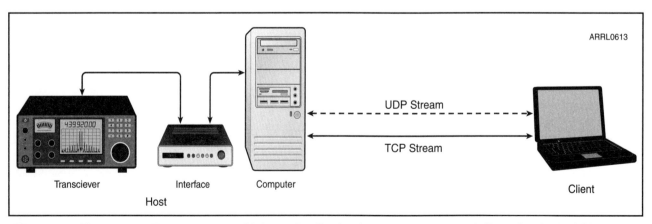

Figure 5.1 – There are really two streams of information flowing between the host and client. One is the TCP data to and from the station hardware. This stream carries commands for the host station and the host station's responses (or it may carry data to emulate the entire desktop of the host computer on the client's screen). The other is a UDP stream of digital audio. This is the transmit audio from the client operator and the received audio from the host radio.

Table 5.1
VoIP Software

Application	Operating System	Source
AOL Instant Messenger	Windows, Linux, Mac OS	www.aim.com/download.adp
BitWise IM	Windows, Linux, Mac OS	www.bitwiseim.com/
BroSix	Windows	www.brosix.com/
Coccinella	Windows, Linux, Mac OS	http://thecoccinella.org/
CounterPath X-Lite	Windows, Linux, Mac OS	www.counterpath.com
EchoLink	Windows, Linux, Mac OS	www.echolink.org
Ekiga	Linux	http://ekiga.org/
Empathy	Linux	http://live.gnome.org/Empathy
eyeP Communicator	Windows Mobile	www.eyepmedia.com
Google Talk	Windows	www.google.com/talk/
iChat	Mac OS	Included in Mac OS
Jabbin	Windows, Linux	http://sourceforge.net/projects/jabbin/
KCall	Linux	www.basyskom.de/index.pl/kcall
KPhone	Linux	http://sourceforge.net/projects/kphone/
Linphone	Windows, Linux	www.linphone.org/index.php/eng
Mumble	Windows, Mac OS, Linux	http://mumble.sourceforge.net/
QuteCom	Windows, Linux, Mac OS	www.qutecom.org/
SightSpeed	Windows, Mac OS	www.sightspeed.com/
Skype	Windows, Linux, Mac OS	www.skype.com/
TeamSpeak	Windows, Linux, Mac OS	www.teamspeak.com/
TokBox	Windows	www.tokbox.com/
Twinkle	Linux	www.xs4all.nl/~mfnboer/twinkle/
Ventrillo	Windows	www.ventrilo.com/
Windows Live Messenger	Windows	http://download.live.com/messenger

Skype is one of the most popular VoIP applications.

weaknesses. However, most hams appear to have settled on a favorite: *Skype*.

Skype debuted in 2003 from Estonian developers Ahti Heinla, Priit Kasesalu and Jaan Tallinn. The *Skype* Group, founded by Niklas Zennstrom of Sweden and Janus Friis of Denmark, has its main headquarters in Luxembourg. *Skype* is popular for several reasons:

■ It is easy to setup and use. It "tunnels" through software and hardware firewalls automatically, so you avoid most of the hassle of port juggling.

■ It supports a wide range of features, including video.

■ It's free

I won't go into detail about how to set up *Skype* because it is so simple to navigate on its own. Once you download and install *Skype*, its setup "wizard" guides you through the steps. However, there are a few things to keep in mind:

(1) To use *Skype* you must set up an account. This means the host station needs an individual *Skype* account, as do all clients. To establish a *Skype* account you must select a User ID and a password. Call signs make excellent User IDs.

(2) As you set up *Skype*, be sure to specify the correct audio devices. For client operators this will be the device driving your headset/microphone. For host stations this will be the sound device connected to the radio.

Look closely at this *Skype* setup screen and you'll see that the box labeled "Automatically adjust microphone settings" has been checked. For our application, particularly at the host station, this box should *not* be checked. We don't want *Skype* to automatically adjust the audio level from the radio (the "microphone audio"). This will cause *Skype* to mute weaker signals that clients will want to hear.

In *Skype*, starting or accepting a call is as easy as a single mouse click.

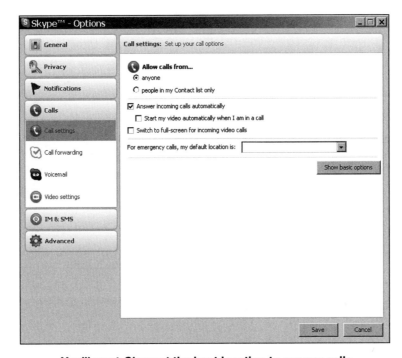

You'll want *Skype* at the host location to answer calls automatically. You'll find this setting under TOOLS, then OPTIONS, then CALLS. Click on the SHOW ADVANCED OPTIONS button. Make sure the box labeled "Answer incoming calls automatically" is checked.

The Level Devil

Many VoIP applications include a feature that adjusts the microphone level automatically. In normal telephone-style usage, this is understandable and even helpful. After all, you want to maintain a proper input level with varying speech patterns. But for Amateur Radio remote control, an automatic microphone level adjustment feature can be a liability.

On the client side, automatic level adjustment can produce some unusual results. If it is too aggressive, it creates a phenomenon where the background noise level rises dramatically whenever the operator pauses. You'll occasionally hear this on the air during normal (non-remote) operating when a ham is using broadcast-style audio processing. When he stops talking, all the background sounds, such as amplifier cooling fans, suddenly increase in volume. It almost sounds like his transceiver is creating a giant vacuum and sucking his station through the microphone!

On the host side, automatic microphone adjustment is more than an annoyance, it is a serious problem. Remember that the audio the software is seeing as its "microphone" input consists of received signals coming from the transceiver. As amateurs it is critical that we hear audio at all levels, including the faint whispers of signals buried in the noise. Chances are, an automatic microphone leveling algorithm will not "hear" those faint signals and won't allow them to be heard by the client operator. Worse still, the software will attempt to "ride" the level as it fluctuates from the radio. The result is a squelch behavior that only provides occasional bursts of audio interspersed with silence.

Fortunately, many VoIP applications allow you to turn off the automatic microphone level adjustment. *Skype* has this option, which is probably another reason why it is so popular among remote station enthusiasts. *Windows Live Messenger*, on the other hand, does *not* allow you to disable automatic level adjustment. This makes it a poor choice for remote operating.

Audio Quality

VoIP bandwidth is relatively narrow by necessity. As a result, audio quality tends to suffer. This effect is most pronounced when the client operator is transmitting. The audio applied to the transceiver microphone jack comes straight from the VoIP application and the sound device. If you're lucky, it won't sound too bad. On the other hand, it may sound strangely flat or tinny. The best transmit audio is the kind that's indistinguishable from that of any other "normal" transceiver on the air. Achieving that ideal isn't always straightforward, however.

Many transceivers offer built-in transmit and receive equalization. If this is true of your host radio, you are indeed fortunate! You can use this feature to easily tweak the transmit audio and boost the low (bass) range, or perhaps reduce the high-end frequencies. The same technique can be applied to the receive audio going back to the client.

If your radio lacks this feature, you can purchase outboard audio equalizers. These devices are not terribly expensive. You can purchase a dual-channel 15-band graphic equalizer from RadioShack for less than $75. You'll find other, less-expensive models from other vendors as well.

Finally, your sound device may have its own equalization software. If so, you can use this application to tailor the transmit audio.

Regardless of which approach you choose, finding the correct equalization settings is a matter of good-old-fashioned experimentation. You may have to sit before the host radio and make adjustments while the client operator solicits comments over the air.

The Telephone: Your Good Old Fashioned Alternative

This chapter has focused on using the Internet as the means for transporting audio signals between the host and client. This is a fine approach for audio streaming from a remote receiver, as we'll discuss in the next chapter.

But making the jump to two-way VoIP increases the complexity. Not only are you using the Internet to allow the client operator to observe and control the radio, you are using the same medium to carry the VoIP audio. An alternative may be to simply use Alexander Graham Bell's favorite invention: the telephone!

There is no reason why you could not employ a dedicated telephone connection to carry audio between the host and client locations. This is not without its own sets of challenges, though. You need to install a device at the host that will answer the incoming call from the client and establish a full-duplex audio connection with the computer sound device, or directly with the radio.

Veteran amateurs will recognize this description as something we used to call a *phone patch*. In this instance, however, we need a device that is more sophisticated than the phone patches of years gone by. The Viking RAD-1 described in Chapter 4 may be suitable. It provides a 600-Ω LINE OUT port for audio access, in addition to its remote control capabilities. Alternatively, some hams have also taken ordinary telephone answering machines and modified them for this purpose. They use the answering machine to "seize" the line when a call comes in, and then patch the two-way audio lines directly to the station equipment.

Another possibility is to "repurpose" an autopatch device otherwise used in an Amateur Radio FM repeater. The Hamtronics AP-3 autopatch (**www.hamtronics.com/autopatch.htm**) will answer incoming calls and set up two-way audio. This was not its intended application, but with some creative repurposing it can work for remote station applications as well.

The advantage of telephone audio is that it tends to be more stable and reliable than VoIP audio. Also, by using telephone audio, you'll reduce the software complexity significantly.

Placing (and Answering) a VoIP Call

The way in which the client connects to the host station will vary depending on the software. In some setups, the client software will combine the station control and VoIP functions, or may automatically run the VoIP software when you attempt a connection.

If your client setup relies on independent control and VoIP applications, it is best to make the control connection first. Make sure you are truly in control of the radio before you start the audio stream.

Placing the actual VoIP call is usually straightforward. Once you've started the software and, presumably, connected to the VoIP network, you're ready to go. *Skype* and similar programs provide contact lists that function as your personal VoIP directory. Just enter the host address once and it will always be available. When you want to place a call, you simply select the HOST ENTER FROM THE LIST and click your mouse on CONNECT or a similar button. You may hear the sound of a telephone ringing. Eventually the host will answer and the audio stream will begin.

Another feature to insist upon in your VoIP software is the ability to answer calls automatically. This is critical to the host station for obvious reasons. With no one available to click the host station's mouse on the ACCEPT? prompt, nothing will happen! Most VoIP applications include this feature, but not all. It pays to shop carefully.

A Few Words about *EchoLink*

EchoLink is a VoIP application designed specifically for Amateur Radio. It was the brainchild of Jonathan Taylor, K1RFD. *EchoLink* is used primarily to link distant FM repeaters and individual stations, but some enterprising individuals have put it to work for remote station control. For now, however, let's spend some time getting to know *EchoLink*.

The program operates in either of two modes — *single-user mode* or *sysop mode*. The basic VoIP functions of both modes are the same, but sysop mode adds features to support a link transceiver or repeater connected to the PC.

One of its key components is a digital audio mixer, which is used to combine audio signals from several sources (audio codec, Morse generator, and speech generator) before sending them to the computer's sound card. *EchoLink* also includes several built-in digital signal-processing (DSP) functions, including a software-based DTMF decoder, which detects and decodes DTMF commands from the link transceiver.

EchoLink has several different input-output (I/O) functions. Of course, the program exchanges digitized voice data with the computer's sound card. But it also includes a user interface, which accepts commands from the keyboard and mouse and displays the program's status graphically. The program exchanges data with the PC's serial port to control the link equipment, and communicates with the Internet to exchange voice and text data with other stations and authentication and status information with the *EchoLink* servers. In sysop mode, a built-in Web server supports remote management of the node, and a program option sends status information to a TNC for distribution over the local APRS packet network.

EchoLink stations communicate directly with each other over the Internet. This kind of communication is called *peer-to-peer* because there is no master — all stations are on equal footing with each other. System engineers are fond of saying that this arrangement *scales* well, because it allows thousands and thousands of QSOs to take place simultaneously, without creating a bottleneck at some central server.

If each *EchoLink* station kept its own list of every other *EchoLink* station it wanted to communicate with, along with a password (or some other form of authentication), no such central server would be required. However, there are many thousands of registered *EchoLink* stations, with more than 170 new ones coming online every day. Most of these stations use Internet providers that

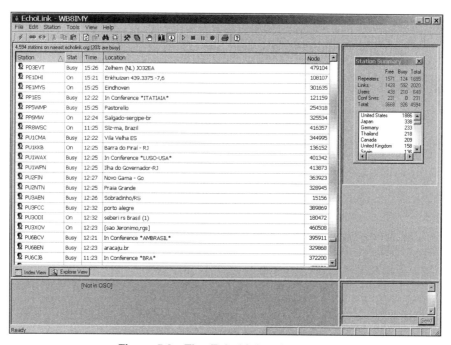

Figure 5.2 – The *EchoLink* main screen.

assign dynamic addresses, which means their Internet addresses change from time to time. Moreover, it's useful for each station to be able to keep running tabs as to who's online at a given moment, rather than simply keeping a big phone book around.

Figure 5.2 shows the main screen that *EchoLink* displays while it's running. The list of stations at the center of the screen is organized by location, using folders and sub-folders similar to *Windows Explorer*. The space bar works like a locking push-to-talk switch — tap once to begin transmitting, tap again to stop transmitting. The horizontal bar graph near the bottom is an audio-level meter that displays both transmitted and received audio levels. Other systems have similar screens and functions in this mode.

A convention adopted by *EchoLink* is to identify RF nodes with a call sign suffix, either -L or -R, and single-user, PC-based nodes with no suffix. The -L suffix indicates that the node is a simplex link, and -R indicates a link to a repeater. On *EchoLink*, even single-user nodes have an assigned node number, so they, too, can be reached via DTMF commands from mobile users working through a distant link.

If you are running a Macintosh computer with the *OS X* operating system, you can also connect to *EchoLink* nodes using a program called *EchoMac*, originally developed by Steven Palm, N9YTY. *EchoMac* is an open-source VoIP package designed to be fully compatible with *EchoLink*, fulfilling the need for desktop Internet linking for Mac users. It can be downloaded from **echomac.sourceforge. net**. *EchoMac* is based partly on a project called *EchoLinux*, originally spearheaded by Jeff Pierce, WD4NMQ, which brings an *EchoLink*-compatible client to the *Linux* platform.

If you're running *Linux* or a newer Mac machine, you might actually be able to run the *EchoLink* software itself. If your Mac has an Intel CPU, you may have already installed a copy of

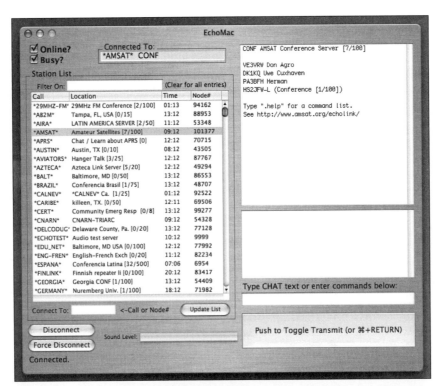

There is also a version of *EchoLink* for Mac users. It is called *EchoMac*. It can be downloaded from echomac.sourceforge.net.

Windows using the *Parallels Desktop* virtual-machine layer. If so, *EchoLink* can be installed in the *Windows* virtual machine just as it would be installed on a PC. If your Mac has either the *Leopard* (10.5x) or *Snow Leopard* (10.6x) versions of *OS X*, you will have a utility called *Boot Camp*. This will allow an Intel processor based Mac to directly run *Windows* natively without the need for any additional software like *Parallels Desktop*. You'll find information on installing *Boot Camp* in your computers documentation.

If you're running on a Mac with an Intel CPU, you may have already installed a copy of *Windows* using the *Parallels* virtual-machine layer. If so, *EchoLink* can be installed in the *Windows* virtual machine just as it would be installed on a PC.

Even if you choose not to use *EchoLink* in your remote-control application, it may be worth exploring as another option for antenna-restricted clients who wish to expand their capabilities. For more information, see the ARRL publication *VoIP: Internet Linking for Radio Amateurs* by Jonathan Taylor, K1RFD.

6

The Listening Post

Portions of this chapter were contributed by Bob Arnold, N2JEU.

If you'd like to start by dipping your toes slowly into the waters of remote station control, one of easiest projects is a remote "listening post." I call it a listening post because the concept reminds me of a Cold War outpost somewhere north of the Arctic Circle where a lonely intelligence officer monitored the airwaves day in and day out. No, I've never served in that capacity in real life, but I knew a few who did. They would have marveled at the remote listening capability we have today!

The receiver at our 21st century listening post can be located anywhere an Internet connection is available and, by the same token, our listeners can be anywhere in the world. They can eavesdrop from the comfort of their living rooms, from gate areas at airports or even from their automobiles if they have mobile broadband.

The applications are limited only by your imagination…

■ A tunable HF receiver

■ An FM receiver eavesdropping on a local repeater

■ A remote receiver monitoring satellite amateur communications

Any receiver that can provide an audio signal to a sound card can be shared over the Internet. I know a retired firefighter who listens to audio from a remote receiver that monitors the fire dispatch frequency in his hometown. Since moving to a community 2000 miles away, eavesdropping on the action was impossible. But thanks to a friend with a scanning receiver and an Internet connection, he can now listen to his old buddies at work any time he desires.

Simple Public Streaming – Step By Step

Let's begin with a project that will allow you to stream the audio from a single receiver to several people at once. See **Figure 6.1**. In this example the receiver is dedicated to monitoring a single frequency. It might be the frequency of a local repeater, or the frequency of a favorite HF net. It's audio output signal is taken from an external speaker jack and fed directly to the LINE or MICROPHONE inputs of a sound card that we'll assume is connected to (or inside of) a dedicated host computer.

The sound card will turn the audio signal into digital data. Now you're ready to stream the data over the Internet. To do so, you'll need two pieces of software. For this example I'll be discussing *Windows* applications since most readers are likely to be using a *Windows*-based computer to host the stream.

First is the piece of software is called *Edcast* (formerly known as *Oddcast*) and should be avail-

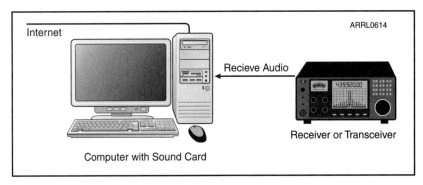

Figure 6.1 – The hardware required for a remote listening post is remarkably simple. All you need is a receiver and transceiver and a computer connected to the Internet. An ordinary shielded audio cable feeds receive audio from the radio to the MIC or LINE input of the computer sound device.

able from **www.oddsock.org**. This is free software.

Since MP3 streams are supported by most operating systems including many versions of *Windows*, Apple *OS X* via iTunes and even many versions of *Linux*, I recommend getting an add-in for *Edcast* to generate MP3 streams. A search for "Lame MP3 encoder" on Google (**www.google.com**) should turn up several sources for your particular operating system including Windows.

When you get the Lame encoder simply copy the file that ends in "DLL" from the Lame installation folder to the folder that *Edcast* installed itself in to enable MP3 encoded streams.

The second piece of software takes the single audio stream from the *Edcast* software and makes it available as multiple copies to whoever asks to listen. It's called *Icecast* and is available in a *Windows* version from **www.icecast.org**. Once

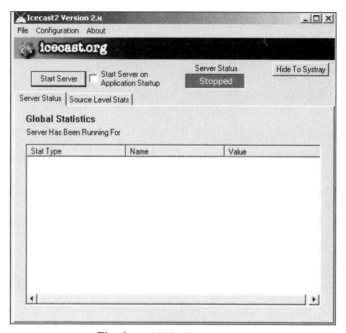

The *Icecast* start-up screen.

If *Icecast* starts up in the "running" mode click on the Stop Server button to stop it, then you can select Configuration then Edit Configuration to bring up a file called *icecast.xml* in a *Windows Notepad* window.

again, this is free software. *Icecast* is available as source code if you need to assemble it for a *Linux* system, although a Web search should turn up already assembled versions for some of the *Linux* variants out there.

Install the *Icecast* software first, then start it up. If it didn't put an icon on your desktop during the install process, go to the folder where it installed itself and start it by double clicking on the EXE file. If *Icecast* starts up in the "running" mode click on the **Stop Server** button to stop it, then you can select **Configuration** then **Edit Configuration** to bring up a file called *icecast.xml* in a *Windows Notepad* window. I suggest making a copy of this file straight away so if you make any errors in changing the configuration you can start again easily.

Find the section that starts with <**icecast**> and within that you'll see a section called <**authentication**>. Change the passwords for the <**source-password**> and <**admin-password**>. Make sure you have these written down elsewhere since you'll need them both for future steps. Save the file and when you click on the button to start the server the changes will take effect.

The other options in the *Icecast* configuration file will enable more features of the server, but what you've done so far are more than adequate. Complete documentation on the configuration file options is available on the *Icecast* Web site. With the additional features you can limit how many streams you can handle at once and how many minutes someone can listen to the streams before they are automatically kicked off.

Next, start up the *Edcast* software. Again, if it didn't add an icon on the desktop go the folder it installed itself in and double click on the EXE file.

When it starts up you'll see a black box that normally shows VU-style meter bars to help you set the audio levels with the slider under the bars. Left click on the black box to turn on the VU meter. The drop-down boxes under the VU meter will let you select the sound card and the input on that sound card that the software will use. Set them appropriately for the sound card and the correct input that you're using. I suggest the line input since it is much more tolerant of input levels than the microphone input.

If you click on **Add Encoder** you'll see a new line appear in the white box. Right click on the line, and then left click on **Configure** in the little box that popped up. You'll now get a screen that lets

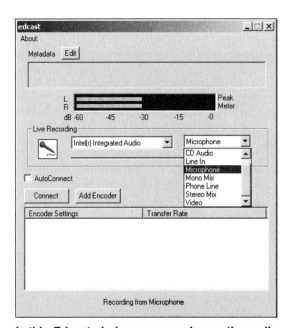

In this *Edcast* window you can choose the audio input. If the audio from your radio is being fed to the MICROPHONE INPUT of your sound card, for example, you would choose this input.

you set up the stream *Edcast* generates. If you've got the Lame MP3 encoder installed properly, you'll see "Lame MP3" in the **encoder type** pull-down box.

For the moment enter these settings:

Bitrate = 16 (the speed of the audio stream in kilobits per second)

Quality is grayed out and not used in mp3 streams

Samplerate = 16000 (the number of bits per second used to sample the incoming audio from the sound card)

Channels = 1 (number of audio channels in the stream.)

Encoder Type = Lame MP3 (should be selected to set the bitrate value)

Server type = Icecast2

Server IP = localhost (leave as-is if the *Icecast* server software is also on the same computer.)

Server port = 8000 (set it to this if it isn't already)

Encoder password = (whatever you set the source password to in the *Icecast* configuration – it's case sensitive!)

Mount point = (whatever you want but it *must* start with a "/". I suggest /test as an example)

Reconnect seconds = 10 (set it to this if it isn't set already – this is the time the *Edcast* software will wait to connect or reconnect to the *Icecast* software if the connection is disturbed.)

Click on the **OK** button in this window to save your settings. I'll discuss the other buttons in that window in a little bit.

Now you should be able to click on the **Connect** button and have the *Edcast* software connect to the *Icecast* software. If the *Edcast* server keeps trying to reconnect, re-check your entries in the configuration settings for that stream. The most likely problem is an incorrect password.

You should be able to point a Web browser on the same computer to **http://localhost:8000** and get a page generated by the *Icecast* software that displays that a stream is available. Clicking on the **m3u** button should start whatever sound utility that handles mp3 streams on your computer (*Windows Media Player*, for example). With luck it will begin playing the stream if you have the receiver turned on and audio feeding into the sound input you've

selected. Other information on the page will give you additional information about the streams.

The Administration link at the top of the page will ask for a name (admin) and a password (use the admin password that you set in the *Icecast* configuration). This provides some additional management utilities that shouldn't be made available to stream listeners, so don't give out that password!

If you want to change several of the things that the basic *Icecast* Web page displays about the stream, just go back to the *Edcast* software and edit the configuration. Using the YP settings button will bring up another box that you can use to set up some information about the stream. The ADVANCED settings button will let you set up *Edcast* to record the stream as a file on the local hard drive simultaneously as it sends the audio over the Internet. I don't recommend this since the files it makes can get very large pretty quickly and run the computer out of storage space. I'd also leave the log stuff in that window alone since it is there basically for diagnosing problems with *Edcast* by the author of the software. Remember to click on the OK button to save the changes if you make any to these options.

Now that you have your audio stream running on your host system, it's time to make it available to the outside world. You may need to make some changes to your router if you're using one in your home network.

Since each router manufacturer does things a bit differently, you'll need to consult your router manual. If you've lost your manual you should be able to visit the manufacturer's Web site and find a copy of the manual as a PDF file in their Support section.

As we touched upon in Chapter 3, it is best to assign a static address in the router for the host computer. That will go a long way toward avoiding a potential conflict with addresses given out by the DHCP server in your router. (If you've forgotten about DHCP and static addresses, re-read Chapter 3.) This is so the computer has the same IP address in your internal network all the time. You may also need to set the gateway, netmask and DNS server values into the computer when you set the static IP on your computer. There should be a screen in the

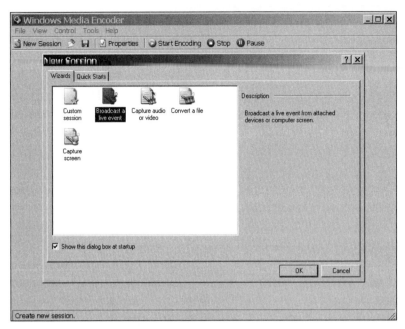

Windows Media Encoder is an alternative for public distribution of receiver audio. This is the *Media Encoder* initial setup screen. Notice the highlighted selection "Broadcast a Live Event." Click this icon and *Media Encoder* will guide you through the process.

configuration settings of your router that displays these for you.

In your router, go into the port forwarding configuration and set port 8000 to go to the static IP address that you're programmed for the host computer. Now any request that is sent to port 8000 from the outside world will be re-directed to the streaming audio host.

Have a friend test your stream by opening his Web browser and entering the following address (we'll assume you're using an Internet IP re-directing service with your call sign [W1XYZ] as part of the address):

http://w1xyz.dynip.com:8000/

If they get the *Icecast* stream display page then the streaming should be good to go. In fact, they should be able to select the stream and listen right away.

There is one small legal point that we should address here. It may occur to some readers that they could receive local AM or FM radio broadcasts and then stream them over the Internet for everyone to hear. It's not a bad idea. Perhaps your hometown friends have moved away and would appreciate listening to their favorite radio station. The problem, however, is that there are laws that govern such an endeavor, especially when it involves making the audio from a broadcast station available to the public via the Internet – even if that "public" only consists of a dozen people. This isn't to say that it can't be done, but you must purchase a licensing agreement and will be billed for each song you play. The costs can soar into the stratosphere very quickly!

The *Edcast/Icecast* approach works well and is very stable, but it is not the only option for public streaming. You can also try MicroSoft's *Windows Media Encoder*. This software package also has the ability to broadcast an audio stream to many listeners at once. *Media Encoder* is a free download at **www.microsoft.com/windows/windowsmedia/ forpros/encoder/default.mspx**.

Single Stream: One Person, One Radio

Let's say you're not interested in public streaming. All you want to do is take the audio from a single, dedicated receiver and make it available to your buddy on the other side of the continent. Maybe he would like to listen to the commuters on the local FM repeater as they travel to and from work. If this is the case, streaming becomes much easier.

The connections between the receiver and your computer sound device remain the same. The difference is that instead of using *Edcast* and *Icecast* for public streaming, we can simply use the free *Skype* VoIP software.

See Chapter 4 for more information about *Skype*. In a nutshell, you simply download, install, establish a *Skype* account and configure. Remember to *disable* the automatic microphone level adjustment feature and *enable* auto-answer. If you've set up *Skype* to select the proper audio device (the one that is capturing the receive audio), your friend should be able to place a *Skype* call to the host and it will answer and begin streaming the receive audio. This setup is astonishingly easy, especially since *Skype* will usually tunnel through your software firewall automatically.

A Tunable Receiver

All of the examples we've discussed so far have assumed that the host receiver was fixed on a single frequency. But there are a number of receivers on the market that offer various forms of computer control. This opens the door to a whole new dimension for your listening post clients. Now they can not only listen to the receive audio, they can control the receiver as well.

You still need to feed the audio output from the receiver to the computer sound device. However, you also need to add the ability for the host computer to control the receiver directly. As we discussed in Chapter 4, this can be accomplished in a variety of ways depending on the type of control port the radio offers. In the example in **Figure 6.2** we're using a "deluxe" sound card interface, a model that includes a CAT (transceiver control) interface to allow the computer to communicate with the radio.

The easiest way for the client to control the host receiver is to use a software application such as *TRX-Manager* (**www.trx-manager.com**). Without describing the software in detail (since the details will likely change anyway), the basic approach requires that the application be installed at both the host and the client. The software on the client side establishes communication with the host and uses it

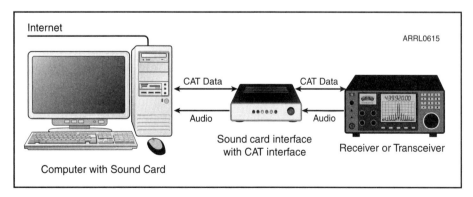

Figure 6.2 – In this example we've added a sound card device that includes a CAT interface. If the receiver or transceiver has computer-control capability, the computer can exercise control through the CAT. This will allow the listener to tune the receiver, switch modes, etc. These interfaces are available from manufacturers such as West Mountain Radio, MFJ, microHAM and so on.

to directly control the receiver, changing frequencies and modes at will. At the same time, the client must also place a *Skype* call to the host so that audio can be streamed as well.

Remember our discussion of ports, firewalls and routers in Chapter 3? Well, to allow the control functions to operate properly, you'll need to open ports so that *TRX-Manager* (or whatever software you are using) can communicate with the client.

Another alternative to consider for receive control is *Virtual Network Computing* (VNC), also known as *remote desktop sharing*. This involves allowing the client operator to actually access the host computer and display its desktop on his machine, just as though he was sitting in front of it at the host location. In this way the client could be in direct control of the host radio software (perhaps a program such as the free *DXLab Commander* at **www.dxlabsuite.com/commander/**). All he has to do is click on the appropriate software buttons to change frequency or mode.

There are many software applications and services that make this possible. One of the most popular is Symantec's *PCAnywhere* (**www.symantec.com/norton/symantec-pcanywhere**). With *PCAnywhere* you install a host application at the host station and a client application wherever the client may be. The client merely double clicks on a *Windows* icon and *PCAnywhere* contacts the host and, within a few seconds, displays the host's desktop view on the client's computer. *PCAnywhere* offers password control and can also restrict clients to certain levels of access.

No doubt you've heard of similar services such as GoToMyPC. *Windows* users probably also know that the *Remote Assistant* function is already built into the operating system. However, it requires that host and client jump through several virtual hoops to establish a connection, so it isn't practical for the type of operation described here.

There is also an application available for the Apple iPhone and iPod Touch known as *Mocha* and several hams have successfully used it to control their stations remotely. *Mocha* is available as an "app" (application) that you can purchase, usually through Apple's *Itunes* store. To use *Mocha*, however, you need a VNC application such as *RealVNC* (**www.realvnc.com**) or *LogMeIn* (**https://secure.logmein.com/US/home.aspx**) running on the host computer.

The downside of all of these desktop-sharing systems is that they require fast computers and fast Internet connections at both ends of the path. This is because the software is displaying the entire host

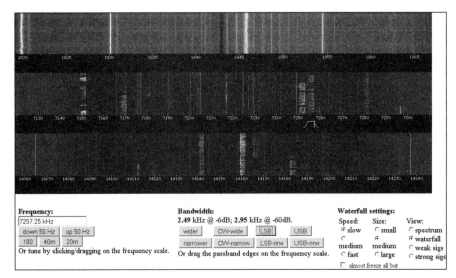

In late 2009 Stan Schretter, W4MQ, set up a Web-controlled receiver site at **http://w4mq.com/** using three Software Defined Radios for 160, 40 and 20 meters.

desktop and "repainting" that desktop display every time something changes. This means you are frequently moving a lot of data between the host and client. Anything less than an optimal setup may result in unacceptable latency for the client. For instance, he may click a software button to change the receiver from SSB to CW and find that it takes several seconds for the change to take effect. And, of course, you still need a VoIP application to port the audio from the host receiver to the client.

Web Receivers

Our discussion of remote listening posts would be incomplete without mentioning Web-based remote receivers. These host receivers are controlled (and heard) through a custom-designed Web page that allows anyone to access the radio. Some Web-based systems require users to establish accounts with user IDs and passwords. Nearly all restrict use to a certain number of minutes so that everyone gets a fair chance to listen to the radio.

One popular site known as GlobalTuners (**www.globaltuners.com/**) aggregates remote receivers from all over the world. GlobalTuners is a free service, but other sites charge user fees.

Setting up a Web-based receiver requires not only an Internet Service Provider to host your Web site, but also the ability to create a functional site using HTML editing tools. A tutorial on setting up a Web-based remote receiver would go well beyond the scope of this book. In fact, it would probably justify a book all its own! But if you'd like to explore the idea, try N2JEU's Web site at **www.ralabs.com/webradio**. Bob Arnold has worked to advance Web-controlled receiver technology and he is constantly developing new tools to streamline the process.

Stan Schretter, W4MQ, has been using Software Defined Radio receivers, particularly the SoftRocks (**http://wb5rvz.com/sdr/**), to create Web-controlled receivers. His work is ongoing and you can learn more at **http://w4mq.com/**.

7

Building a Complete Remote Station

In Chapter 6 we discussed some ideas for setting up remote receiving stations. Now it's time to take the next step: remote transceiving.

Obviously, when you make the transition from *listening* to RF to actually *generating* RF, your list of options (and challenges) expands quite a bit. This opens the door to some exciting possibilities, the range of which are limited strictly by your imagination.

Software Choices

When you're talking about a simple remote receiver, as we were in the previous chapter, your software concerns are relatively simple. Now we must venture into more complex software that supports two-way communication for true user control.

One of the most efficient ways to accomplish Internet remote control is to exchange the least amount of data possible between the host and the client. At minimum, the client software needs to communicate commands to the host, such as changing frequencies or switching from receive to transmit. At the same time, the host must send brief bursts of information about the state of the radio (and the rest of the station) to the client. By keeping these exchanges brief and the required throughput low, it is possible to do remote station control using slower Internet connections.

A less efficient, but more flexible option is to essentially allow the client to "take over" the host computer and operate as though he were seated in front of it. The downside of this approach is that it is much more demanding of your Internet resources. High-speed connections at both ends of the "pipe" are necessary.

Of course, regardless of how the client is controlling the host, there is also the issue of transmit and receive audio, which we addressed in Chapter 5. Depending on the approach you choose, this may be handled by separate VoIP software (such as *Skype* **www.skype.com**) running in tandem with your station control application.

Low Throughput Control

Let's look at several software options for remote station control using the "low throughput" method.

The *Internet Remote Toolkit* for *Windows*, or *IRT*, is the brainchild of Stan Schretter, W4MQ.

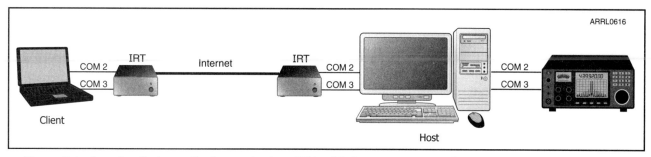

Figure 7.1 – Imagine that our client operator has *IRT* on his laptop computer along with a copy of *MixW*. IRT is running on the host station computer as well. The client operator first starts *IRT* and makes the Internet connection to the host. Once the connection is established, the client operator fires up his *MixW* software. Immediately *MixW* uses COM 2 to poll the transceiver and determine its operating frequency. *IRT* automatically transports the poll request intended for COM 2 over the Internet to the host computer where it reaches the radio via the "real" COM 2. The transceiver "replies" and this data flows back to COM 2 on the host computer, back over the Internet to the *virtual* COM 2 via *IRT*, and finally to the client's *MixW*.

The techniques vary, but in each case the idea is to exchange host and client data as efficiently as possible.

Internet Remote Toolkit

The *Internet Remote Toolkit for Windows*, or *IRT*, is the brainchild of Stan Schretter, W4MQ. It is free of charge and available for downloading at **www.w4mq.com/indextop.html** (look for the **DOWNLOAD** link).

IRT provides an Internet-based control link between software on a client's computer and the physical hardware attached to the host computer. Unlike some remote control applications, *IRT* does not attempt to emulate a transceiver on the client's computer monitor. Instead, *IRT* is designed to create a seamless method for the client to run his desired software and have it operate as though it was directly connected to the actual equipment at the host station. A VoIP application known as *IP-Sound* is included to provide full duplex audio capable of supporting all modes of operation, including digital modes.

Let's illustrate how *IRT* works with a common example. Refer to **Figure 7.1**. *MixW* is a popular *Windows* program that allows hams to operate a variety of digital modes such as RTTY and PSK31. Normally you'd configure *MixW* to use particular COM ports on your station computer and, through a sound card interface connected to those ports, key your transceiver and even change its frequency. *MixW* uses one COM port for CAT functionality to control the transceiver's frequency and mode settings. It uses another COM port to switch the transceiver between transmit and receive, and still another for FSK RTTY keying. For the sake of discussion, let's assign fictional COM ports to all three functions…

- COM 2 = CAT
- COM 3 = Transceiver switching
- COM 4 = FSK

Imagine that our client operator has *IRT* on his laptop computer along with a copy of *MixW*. His *MixW* has been configured with the COM port assignments above. In the meantime, *IRT* is running at the host location, waiting for someone to make a connection.

The client operator first starts *IRT* and makes the Internet connection to the host. Once the connection is established, the client operator fires up his *MixW* software. Immediately *MixW* uses COM 2 to poll the transceiver. *IRT* automatically transports the poll request intended for COM 2 over the Internet to the host computer where it reaches the radio via the "real" COM 2 on the host computer. The transceiver "replies" and this data flows back to COM 2 on the host computer, back over the Internet to the *virtual* COM 2 via *IRT*, and finally to the client's *MixW*.

Let's say the client operator has placed *MixW* in the PSK31 mode. *IRT* is sending audio back from the host radio, so the client sees the PSK31 signals on his *MixW* waterfall display. He can click his mouse on any signal and decode the text directly.

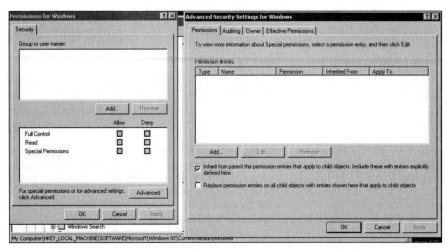

Hamachi VPN "tunnels" through all security gateways directly between the client and host computers, thus providing the end-to-end remote connectivity without the hassles of network and computer security.

When he wants to transmit, he clicks his mouse cursor on the *MixW* **TRANSMIT** button. *MixW* sends the transmit keying signal through COM 3 to *IRT*, which relays it to COM 3 on the host computer and onward to the transceiver. The transceiver enters the transmit mode and the PSK31 transmit audio from the client's *MixW* software flows through the *IRT* VoIP link to the host computer, exiting the sound device and directly into the transceiver microphone jack.

As far as *MixW* is concerned, it is talking to the transceiver directly. Thanks to *IRT*, all the COM port signaling and audio shuffling is completely invisible to the client operator. When it works well, the illusion is nearly perfect. You could sit someone before the client computer and tell them the computer was connected to a transceiver in the next room…and they'd believe it!

IRT makes it possible to use just about any software you wish at the client location. All you need to do is configure all the COM ports at the client and host to "match," then use *IRT* to bridge the gap between them via the Internet.

For convenience, the *IRT* also provides some built-in user interfaces: radio control for Kenwood radios, Alpha87 amplifier control, SteppIR antenna control, a DXcluster interface, a rotator interface, a CW keyboard and a logging interface. These are provided for initial ease of startup (at least with Kenwood radios), but may be easily replaced by any other software: ham radio developed, commercial or home brew.

The need for layers of security both within computers (e.g. firewalls) and within network access nodes (e.g. routers) can complicate the end-to-end functioning of a remote even for the most computer savvy individual. That's why this book spends so many pages talking about ports and other arcane topics. But if you are simply setting up a single host and a client that need to access each other, you may wish to consider using the *Hamachi Virtual Private Network* (*VPN*) software (**https://secure.logmein.com/products/hamachi2/**) to effectively "tunnel" through all security gateways between the client and host computers, thus providing the end-to-end remote connectivity without the hassles of network and computer security. Since this VPN is accessible only through password access established by the user, security access to the computers is still maintained.

Of course, if you are not using VPN, you'll need to make sure the necessary firewall ports are open at the client and host, and that the router (if you are using one) is forwarding these ports between the respective computers.

COM Port Redirectors

Depending on how knowledgeable you are about computer networking, it may have occurred to you that *IRT* sounds an awful lot like an Amateur Radio version of a *COM Port Redirector* or *TCP Bridge*. And you'd be right!

A COM Port Redirector is a piece of software that re-directs information intended for a local COM port to a remote COM port at another location, often via the Internet. Our previous *MixW* example would work just as well if *MixW* were sending its data through commercial COM Port Redirectors at both the host and client locations. That is to say, if you have access to COM Port Redirecting software, you don't need *IRT*. In addition to the redirecting application, all you would need would be VoIP software such as *Skype* to transport the audio.

The catch, however, is that COM Port Redirectors can be expensive, whereas *IRT* is free. For example, *NetSerial* by PCMicro (**http://pcmicro.com/netserial/**) sold for $90 per computer when this book was written. You need *two* licenses, one for the client and one for the host, for a total investment of $180. If your remote setup consists of a single host and a single client, $180 may not be all that much money in the grand scheme of things, but it makes the $0 price tag of *IRT* look awfully attractive.

As I said at the beginning of this book, times have a tendency to change. So, with some careful research, you may stumble upon free or low-cost redirectors. This is especially true if your host and client computers are *Linux* machines since most *Linux* software is free.

We're also beginning to see the appearance of special Ethernet adapters that are designed to make any computer-controlled transceiver a network device that can be controlled directly over the Internet. As you'll see later in this chapter, at least one of these devices offers built-in COM Port Redirectors.

Internet Remote Base

Internet Remote Base, or *IRB*, was originally written by Stan Schretter, W4MQ and is now maintained by Bob Arnold, N2JEU. This *Windows* application is free for downloading at **www.n2jeu.net** (look for the link to the **Downloads** page).

IRB is a sophisticated package that consists of two applications: *WebXCVR*, the client software, and *IRBHost*, the host software. *IRB* provides transceiver emulation – a screen that resembles a transceiver – for the client operator. It also contains

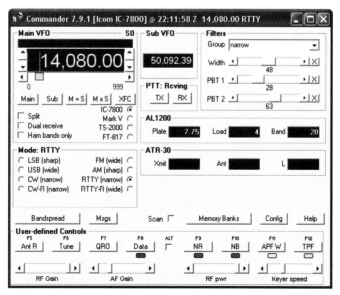

IRBHost can work indirectly with most other computer-enabled radios via the DXLab Commander program by Dave Bernstein, AA6YQ, at www.dxlabsuite.com/commander/.

its own VoIP audio application, although you could just as easily use your own.

At the time this book was written, *IRBHost* was designed to work directly with Kenwood TS-2000, TS-480 and TS-570 transceivers. It could also work indirectly with most other computer-enabled radios via the *DXLab Commander* program by Dave Bernstein, AA6YQ, at **www.dxlabsuite.com/commander/**. *Commander* provides a "generic" transceiver emulation that will function with many different transceivers.

The *IRB* software is targeted more for group and club use since it has the ability to store many different login names and passwords, although single-user use is possible. Each user account on the remote can be restricted by license class in the host software, so the capabilities presented to each user could be different. Some could have access to an external amplifier while, for example, the HF amplifier could be disabled automatically for Technician class users. Specific antennas could be made available to users depending on the bands the antennas are capable of handling. Proper configuration could prevent a user from trying to push 1000 W into an antenna that isn't capable of handling the power, or prevent a user from selecting an antenna if it isn't designed for the desired band.

TRX-Manager

TRX-Manager is a *Windows* application by Laurent LaBourie, F6DEX, that approaches the issue of remote station control in a somewhat different manner.

Like *DXLab Commander*, *TRX-Manager* creates a user interface that is an approximation of the transceiver model you've selected (it supports a large number of transceivers). *TRX-Manager* is also capable of controlling antenna switches and rotators. The software does not include its own VoIP application for remote audio, but it can automatically start the application of your choice.

There are no host or client versions of *TRX-Manager*; the same software must be running at both locations. What makes *TRX-Manager* different from other remote-control applications is the fact that it exchanges data between the host and client over the Internet using a *Telnet* connection.

Telnet (Teletype Network) is an "old" network protocol that predates the Internet itself. Without

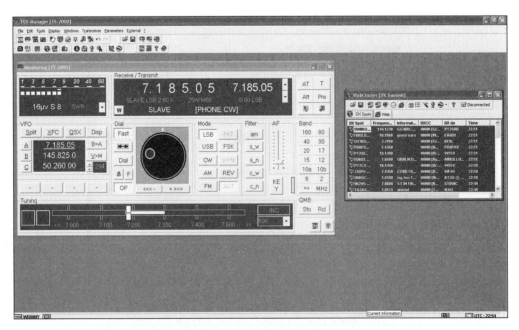

TRX-Manager is a popular remote control application that supports a number of different radios and also offers antenna switching and rotator control.

going into unnecessary detail, Telnet was used to provide an interactive link between a host and client, usually with text as the primary mode of communication. Although Telnet is rarely used in Internet applications today, it still exists as an active protocol, which means it can still be used and recognized.

TRX-Manager uses Telnet for a good reason: unlike applications that require a range of TCP ports to be opened, Telnet requires only one port, typically Port 23. Many networks do not block Telnet ports, so this makes it somewhat easier to get a *TRX-Manager*-based remote-control system running successfully. However, you still may need to make "exceptions" for *TRX-Manager* in your software firewalls at the host and client ends.

When the client makes a change, such as a request to change bands on the host transceiver, *TRX-Manager* responds by sending a string of text to its corresponding *TRX-Manager* application at the host location. The host *TRX-Manager* then translates the text into command data to be sent to the radio. The same process works in reverse.

TRX-Manager is distributed in the US and Canada by Personal Database Applications at **www.hosenose.com/trx-manager/**. The selling price is $79.

Ham Radio Deluxe

Ham Radio Deluxe is a popular free application for *Windows* that supports a wide variety of functions: logging, DXcluster access, mapping, digital communication and much more. Among its many features is so-called client/server functionality.

In a remote-control application, *Ham Radio Deluxe* runs on the host computer with its server mode active, waiting for a connection from the client. The client must also be using *Ham Radio Deluxe*, but with its client function enabled. When the client operator requests a connection to the host over the Internet, the server answers and immediately begins sending and receiving data. Now the client operator is able to use *Ham Radio Deluxe* to control the transceiver.

As with *TRX-Manager*, however, you must supply a separate connection for audio, usually a VoIP application of your choice. You can download *Ham Radio Deluxe* at **www.ham-radio-deluxe.com**.

EchoLink and ARTiE

EchoLink is a popular VoIP network dedicated to Amateur Radio use. Developed by Jonathan Taylor, K1RFD, *EchoLink* uses the Internet to create audio links between FM repeater systems and individual stations.

Len Stefanelli, N8AD, has exploited *EchoLink*'s functionality in a creative way. His free *ART-DX*

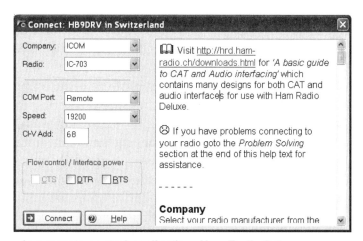

In a remote-control application, *Ham Radio Deluxe* runs on the host computer with its server mode active, waiting for a connection from the client. The client must also be using *Ham Radio Deluxe*, but with its client function enabled.

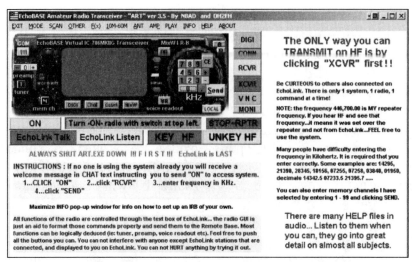

Len Stefanelli, N8AD, has exploited *EchoLink's* functionality in a creative way. His free *ART-DX* host application for *Windows* works with *MixW* multimode software to allow transceiver control via *EchoLink*. The *MixW* software acts as the rig control application while *ART-DX* serves as the "bridge" between *EchoLink* and *MixW*. On the client side, all that's needed is *EchoLink* software along with the client *ARTiE* program (shown here).

host application for *Windows* works with *MixW* multimode software to allow transceiver control via *EchoLink*. The *MixW* software acts as the rig control application while *ART-DX* serves as the "bridge" between *EchoLink* and *MixW*.

On the client side, all that's needed is *EchoLink* software along with the client *ARTiE* program. Each time the client clicks his mouse cursor on a button in the *ARTiE* screen, which emulates the front panel of a generic transceiver, the command is sent automatically as a text string to the host via *EchoLink's* "chat" function. This is similar to the technique used by *TRX-Manager*. It is worth noting that you can also use *ARTiE* with *Skype* rather than *EchoLink* to carry the host/client audio.

ART-DX is free of charge and is available at **www.hfremote.us/ART-DX.html**. *EchoLink* is also free and can be downloaded at **www.echolink.org/**. *MixW* is available for $50 at **http://mixw.net/**.

Virtual Network Computing

Virtual Network Computing (VNC) is a fancy term for "remote control." It is a method of making the desktop of one computer visible – and controllable – on another computer over the Internet.

All the remote control software options I've described so far were designed in the days when high-speed Internet connections weren't commonplace. As we discussed at the beginning of this chapter, these applications held fast to the idea of swapping as little data as possible. If your Internet connection is slow, this makes perfect sense, but today high-speed Internet is no longer a rare luxury. At the time of this writing, roughly 70% of US households were blessed with high-speed "broadband" access. We can expect this percentage to reach 90% in the coming years. In addition, the access speeds will only increase.

The expanding availability of broadband is changing the Internet remote-control landscape. Hams are gradually moving away from the types of host/client applications we've been describing thus far. Instead, they are using Virtual Network Computing tools to simply reproduce the appearance and function of the host station computer on the client's monitor. It is as though the client is seated directly in front of the host

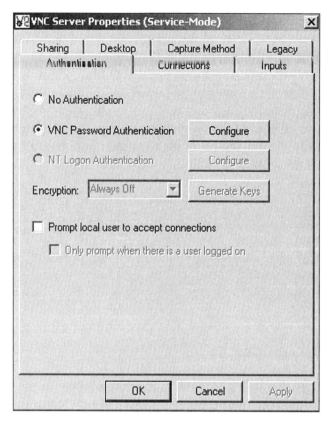

The trend in remote station control is Virtual Network Computing (*VNC*) with applications such as *RealVNC*. Other than a VoIP application such as *Skype* to carry the audio, the client only needs a copy of the VNC program on his computer. At the host computer, all that's needed is another copy of the VNC software and a program that can control the radio and any other attached devices such as antenna switches or rotators. That program could be *DXLab Commander*, *Ham Radio Deluxe*, *TRX-Manager* (not operating in its remote-control mode) or any other station or transceiver control program that suits your fancy. If it can run on the host computer, it can be controlled by VNC.

computer. VNC hasn't been an attractive option in the past because you needed a speedy connection to make desktop sharing possible without mind-numbing delays. But now that the need for speed is being met across the country, VNC is moving to the forefront of remote station control.

One advantage of VNC is operational simplicity. Other than a VoIP application such as *Skype* to carry the audio (yes, you still need that), the client only needs a copy of the VNC program on his computer. At the host computer, all that's required is another copy of the VNC software and a program that can control the radio and any other attached devices such as antenna switches or rotators. That program could be *DXLab Commander*, *Ham Radio Deluxe*, *TRX-Manager* (not operating in its remote-control mode) or any other station or transceiver control program that suits your fancy. If it can run on the host computer, it can be controlled by VNC.

Another advantage of VNC is that it is often *platform independent*, which is just another way of saying that the host and client computers don't necessarily have to be running the same operating systems. A client with a Mac computer, for example, can use VNC to remotely access a *Windows* computer.

There are two principal *disadvantages* to using

Building a Complete Remote Station 7-9

A Remote Station for Digital Modes

If you enjoy HF digital modes such as PSK31 and others, many of the station configurations in this chapter can be adapted to this type of operating as well.

If you can implement COM Port Redirectors at both ends, the client could use these, along with his favorite sound-card-based digital application. Transmit/receive keying would be relayed through the virtual COM ports while transmit/receive audio would travel via the VoIP application, just like voice signals. This would allow the client to see the digital signals on his software display, just as though he was present at the host station.

The alternative is to use a VNC application and run the digital software at the host computer. If the host station has an Internet upload speed of 650 kbps or greater, this may work reasonably well. Otherwise, the host will be tasked with sending a constantly changing desktop (because of the fluctuating software display) back to the client with a considerable delay, probably on the order of two seconds or more. That can be frustrating for the client since he'll be typing his text into the transmit buffer without an immediate echo. If he makes a mistake, he won't know it until he has typed the entire sentence (assuming he is a fast typist).

VNC for remote station control. If the host or client computers are using asymmetrical Internet access with relatively slow upload speeds (less than 650 kbps), latency can become a problem. The client operator may experience, for example, a delay of a second or two between the time he clicks his mouse cursor on a button to select an antenna and the time he sees that switching has actually taken place. If the host desktop display is "busy," say with a continuously scrolling waterfall showing PSK31 signals, this effect can be particularly pronounced. See the sidebar, "A Remote Station for Digital Modes."

The other issue involves security. With VNC the client operator has complete access to the host computer. Unless the control operator configures the host computer to deny access to certain areas, this could be a sensitive issue. Fortunately, most VNC applications provide ways to limit how much of the host machine the client can access. And as we discussed earlier in the book, this is also a strong argument for using a dedicated computer at the host location, a computer that doesn't contain sensitive information.

Of course, with VNC there is still the problem of firewalls and routers. You'll need to configure both to allow the VNC applications to communicate. The VNC documentation usually specifies which ports you need to open and some applications such as the *LogMeIn* software use Virtual Private Networking to keep port hassles to a minimum. Remember that you must have VNC software at the remote (client) and host locations. The client VNC only operates when the client needs it, but the host VNC must run continuously, forever standing by for a connection.

A quick Web search will show that there are many VNC applications available. Some are free, some are not. The features they offer vary widely. Popular VNC packages include *TightVNC* (**www.tightvnc.com**), Symantec's *PCAnywhere* (**www.symantec.com/norton/symantec-pcanywhere**) and *RealVNC* (**www.realvnc.com/vnc/index.html**), just to name a few. There are also VNC-style services that use third-party computers and Virtual Private Networking to implement remote control, such as GoToMyPC and LogMeIn (**https://secure.logmein.com/US/home.aspx**).

It is even possible to perform VNC control over cellular or WiFi networks from your Apple iPhone, iPod Touch or other "smartphone" with applications such as *Mocha* (**www.mochasoft.dk/iphone_vnc.htm**). *Mocha* is not VNC software by itself; it is a client application. You still need a VNC host application from *RealVNC* or another provider to run on the host computer. You will also need *Skype* running at both ends to carry the audio.

Public vs Private

Most ham stations operated by Internet remote control are private affairs. That is to say, access is limited to just a few individuals. Everyone in the group knows the station's Internet address, as well as the user ID and password required by the control software. They coordinate among themselves and establish schedules to determine who can use the station, and when.

In some instances access is very private – the remote station is used by just one operator. I know of an apartment-dwelling amateur who set up a station at his parent's home 40 miles away. The station is strictly for his use.

A few hams have established public stations. These stations are a bit like public Web cameras you'll encounter on the Internet – anyone can use them at any time. A number of public stations use *EchoLink* for audio linking and control, along with the *ARTiE* software created by N8AD. Creating and managing a public station is a larger challenge, but can be rewarding. Keep in mind, however, that the FCC will hold you responsible for whatever takes place!

Station Configurations

Remote-control systems can be as simple or complex as you desire. It really boils down to how much time and money you wish to spend. Let's consider a few ideas. Note that each configuration also includes the telephone-based "emergency shutdown" system we discussed earlier in the book.

#1 – The Simple Station

The station shown in **Figure 7.2** gives the client access to voice operation on a single band. A single band, single mode setup is relatively easy to implement because you don't need to use antenna tuners or switches. Although the client is limited to transmitting on one band, nothing prevents him from listening on several bands, even though the antenna may be less than optimal on a given frequency.

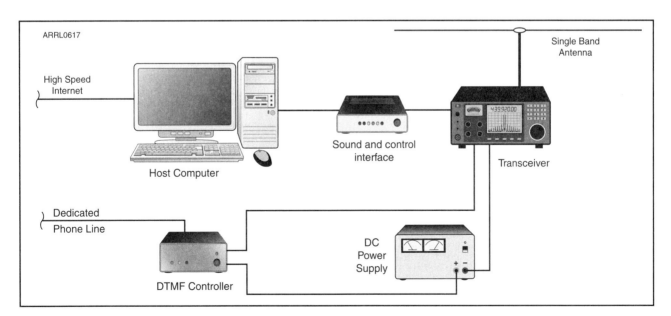

Figure 7.2 – Here is an example of the simplest remote-control approach possible. The single transceiver is interfaced to the host computer. It is also attached to a single-band antenna (no antenna switches or tuners necessary). Notice that this example includes the DTMF "failsafe" shutdown system we discussed earlier in the book. To refresh your memory, its purpose is to give you an alternate means of shutting down the transceiver . . . just in case.

#2 – Multiband with a Single Antenna

In **Figure 7.3** we've gone from a single to a multiband antenna. To keep it simple, however, it is critical that the antenna not require the use of a tuner to operate on its individual bands. In this example we're using a multiband trap dipole that provides a low SWR on every band.

It is worth mentioning that some remotely tuned antennas such as the SteppIR models provide control units that can operate automatically by tapping into a transceiver's band data output, or the host computer's serial port. These antennas can adjust themselves automatically when the client switches bands, or the client may be able to initiate the adjustments himself.

Another possibility is a single antenna with an RF-sensing antenna tuner that automatically seeks the lowest SWR for a given band whenever the client keys the transceiver.

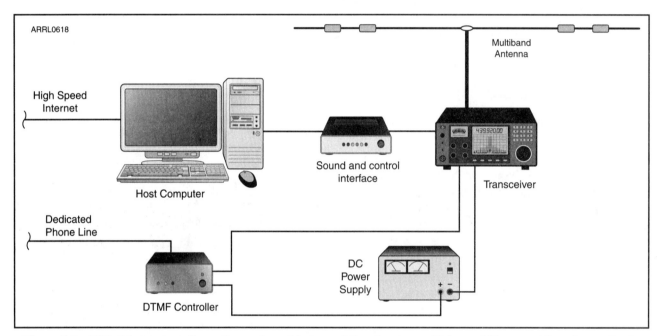

Figure 7.3 – With this host configuration we've added a multiband antenna. This could be a multiband trap dipole of any other type of antenna that switches bands automatically (perhaps through an RF-sensing remote antenna tuner). Simplicity remains the key attribute. The host computer controls the transceiver through a CAT interface and sends/receives audio and does transmit/receive switching through the same interface. Alternatively, the host computer may communicate with the radio through separate CAT and audio/switching interfaces.

#3 – Multiple Bands and Multiple Antennas

Remote switching between multiple antennas is not that difficult. There are antenna switches available that connect to computer serial ports for remote control (**Figure 7.4**), and some provide their own control programs. Others will connect to so-called *band decoders* that select the proper antenna whenever the transceiver changes bands. You'll even encounter switches that will connect directly to a transceiver's band-data output.

See the advertising pages of *QST* magazine for remote-control antenna switches manufactured by MFJ Enterprises, DX Engineering (**www.dxengineering.com**) and Array Solutions (**www.arraysolutions.com**), among others.

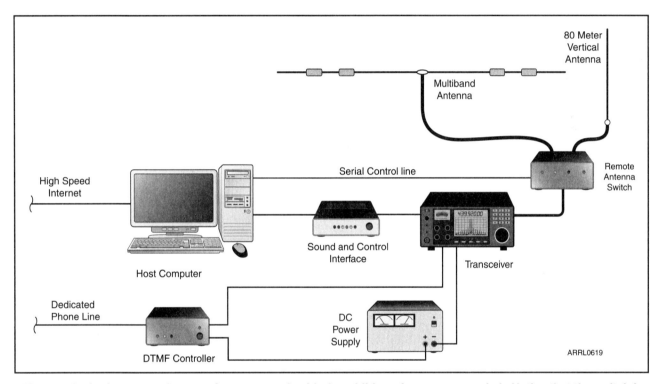

Figure 7.4 – Let's turn our host station up a notch with the addition of an antenna switch. Notice that the switch is connected directly to the host computer. It is assumed that the computer will be running a host application that supports antenna switching (most likely through one of the computer COM ports), or the computer may be running a separate antenna switching application. The antenna switch itself can be located indoors or outdoors.

#4 – Adding an Antenna Rotator

In **Figure 7.5** we have two antennas at the host station, one of which uses a rotator. As we discussed earlier in the book, many antenna rotators can be controlled remotely, often through third-party devices that connect between the rotator control hardware and the computer. Some remote station control applications such as *TRX-Manager* have the built-in ability to "talk" to these devices. The trick is to communicate the antenna direction back to the client operator. Remote control by VNC comes in handy in this application since the client can simply bring up the host computer's rotator control software directly and adjust the position of the antenna.

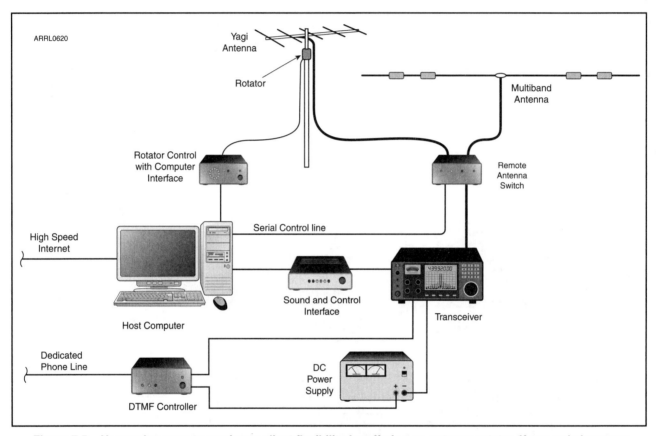

Figure 7.5 – Now we've gone to maximum client flexibility by offering an antenna rotator. If you only have one antenna, the antenna switch is obviously unnecessary. However, you will need a device to interface the rotator controller to the computer. We discussed two such devices earlier in this book. Also, the host application must support rotator control (along with the means of passing the antenna position information back to the client), or you'll have to run a separate application to "talk" to the rotator control interface.

#5 – Remote CW

Yes, CW by Internet remote control is possible, but success varies. The easiest way is to use VNC and allow the client operator to bring up a piece of host software – either a CW/digital multimode program such as *Fldigi* (**www.w1hkj.com/Fldigi.html**), or a dedicated CW terminal such as *CWType* (**www.dxsoft.com/en/products/cwtype/**). The client would simply type the text and the software would translate it to Morse code and key the radio accordingly. The CW receive audio would go back to the client via the VoIP link (**Figure 7.6**).

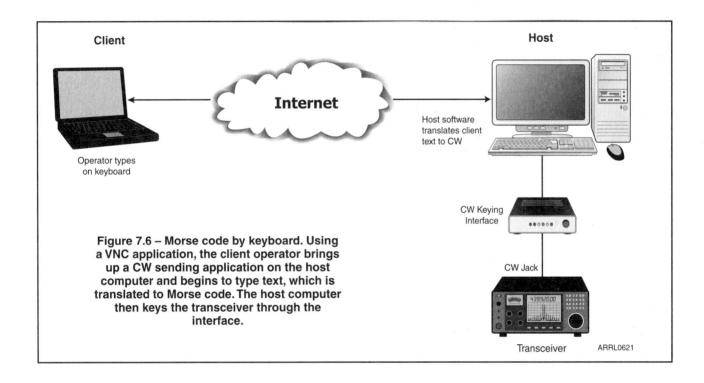

Figure 7.6 – Morse code by keyboard. Using a VNC application, the client operator brings up a CW sending application on the host computer and begins to type text, which is translated to Morse code. The host computer then keys the transceiver through the interface.

CW purists may balk at this idea. It is really little more than digital keyboarding, not far removed from operating RTTY or PSK31. K1EL offers an alternative that brings real CW keying back into the picture. The solution is a bit of software called *WinKeyer Remote Control* that allows two of his WinKeyer CW keyers to transfer their on/off keying outputs to each other – one at the host and one at the client – over the Internet (**Figure 7.7**). The software is free and available for download at **http://k1el.tripod.com/WKremote.html**.

WinKeyer Remote Control allows two of K1EL WinKeyers to transfer their on/off keying outputs to each other – one at the host and one at the client – over the Internet.

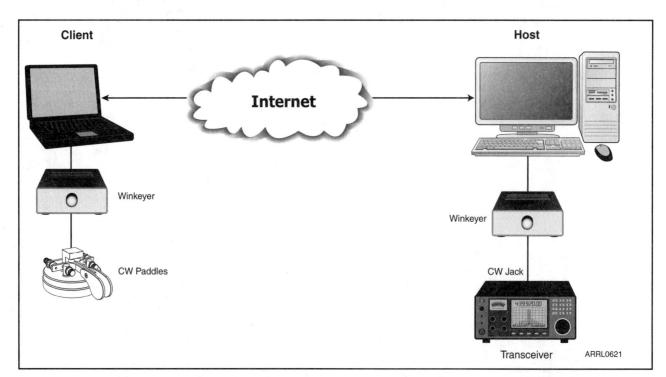

Figure 7.7 – "Real" CW by remote control. Both the client and host computers are running *WinKeyer Remote Control.* The client operator works a set of CW paddles attached to a K1EL WinKeyer, which is itself attached to the client computer. The on/off keying is translated to data and transferred to the host computer and, finally, to the host WinKeyer that keys the radio.

Adding Video

A number of remote-control stations, especially private systems, have added video to enhance the client experience. This is more than just a clever toy; video has very practical applications. And since many VoIP programs also support video conferencing, it is relatively easy to implement.

The video camera need not be expensive. A simple USB video camera, often referred to as a "Web cam," can be yours for less than $20 on auction sites such as eBay (**www.ebay.com**). You're not looking for stunning Technicolor or high resolution, so any of these bargain-basement cameras are more than adequate. The cameras sport USB cables that plug into the host computer directly, but these cables may be rather short since many of the cameras are intended for use with laptop computers. For desktop computers, look for a camera with at least a 6-foot cable, or purchase a USB extension cable.

Also look for a video camera that does not include a built-in microphone. Why? If the camera has its own microphone, your VoIP software may attempt to use it exclusively for the audio stream going to the client. This behavior makes perfect sense for the average (consumer) user, but we want to stream the transceiver audio to our clients, not the camera microphone audio. Besides, a built-in microphone just adds to the cost of the camera.

So what views will you share with your clients? Some host cameras are simply aimed at the primary transceiver to give the operator an alternative means of seeing what the rig is doing. I know of one host station that aims its video camera out a nearby window to allow the client to see the antenna tower. When the client operator commands the rotator to change direction, he can see the antenna moving to confirm the fact!

If you're considering video for your host station, keep in mind that streaming video is a serious Internet bandwidth hog. I don't recommend video unless both the host and client enjoy high-speed connections (650+ kbps upstream at the host; at least 1.5 mbps downstream at the client). Otherwise, the client will be treated to a view that is jerky and fragmented. The video data overload may also seriously impact the audio stream.

#6 – No Host Computer Required!

In 2009 GlenTek (**www.glentekcorp.com**) began selling Ethernet devices known as *RTEs* (Radio-To-Ethernet) that make it possible for any computer-capable transceiver to be network aware and controlled remotely. RTEs plug into transceiver computer control ports, but they are not mere Ethernet adapters. In addition to turning the transceiver into a network device (complete with its own IP address, just like a computer, printer, etc), the GlenTek RTEs also provide virtual COM port redirection as we discussed earlier in this chapter. This means that a client operator could run rig-control software on his computer with COM port signals sent over the Internet and routed directly to the radio. Once again, this setup (see **Figure 7.8**) would appear to the client as a locally controlled radio. No computer is necessary at the host location since the RTE is doing all the communication and translation.

However, there is a small catch. As always, audio is the exception. If you want to send audio to and from the radio, you still need a VoIP application running on a host computer. There may be some creative ways around this problem, such as using a telephone link to transport the audio in both directions. In that scenario, the client operator would literally place a telephone call to establish the audio connection, and then access the transceiver control via the Internet. Also, in early 2010, GlenTek began offering units that were capable of transporting audio as well.

As I mentioned previously, the use of Ethernet ports is slowly increasing among Amateur Radio transceiver manufacturers. For radios that already offer this feature, you'd obviously not need an RTE.

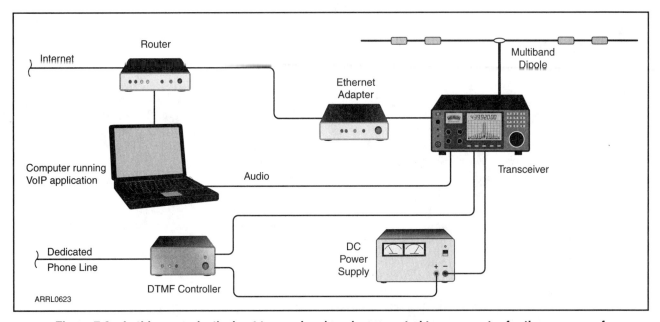

Figure 7-8 – In this example, the host transceiver is only connected to a computer for the purpose of transferring audio. The actual control is taking place directly through an Ethernet adapter. Since the Ethernet adapter and the host computer are both attached to the router, both share the Internet connection and are assigned separate IP addresses. If you had another means of transferring audio, such as a telephone connection or an adaptor that handled control *and* audio (such as the new GlenTek and SM20 units), you wouldn't need the host computer at all.

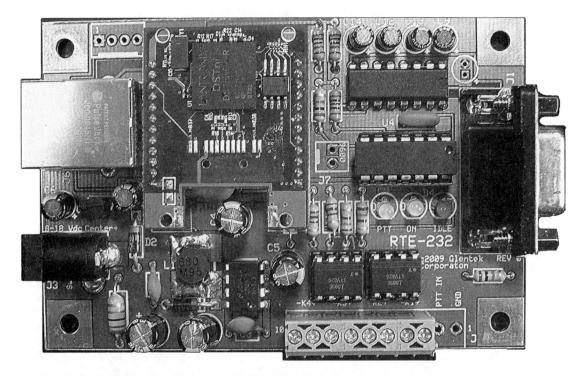

The GlenTek *RTEs* (Radio-To-Ethernet) make it possible for any computer-capable transceiver to be network aware and controlled remotely. RTEs plug into transceiver computer control ports, but they are not mere Ethernet adapters. In addition to turning the transceiver into a device on the network (complete with its own IP address, just like a computer, printer, etc), the GlenTek RTEs also provide virtual COM port redirection.

SM2O's RRC-1258MkII package consists of two devices: one placed at the host and the other at the client. These units not only handle remote control functions, but audio as well.

In early 2010, Mikael Styrefors, SM2O, developed a device known as the RRC-1258MkII. This is a completely self-contained "turnkey" solution for remote control of several types of ICOM, Kenwood, Yaesu and Elecraft transceivers. The RRC-1258MkII is actually sold as a package of two devices: one placed at the host transceiver and the other at the client location. They handle not only the command and control functions, but also the audio stream. Therefore, a computer at the host location is unnecessary unless you need it to control other items such as antenna rotators. The disadvantage, on the other hand, is that every client would need to have an RRC-1258. If you're putting together a remote station that is intended for a single client, this isn't an issue, but it would be problematic for multiple users. When this book went to press, the RRC-1258MkII package was selling for $450 and could be ordered online at **www.remoterig.com/en/**.

Power Output and SWR Monitoring

When you operate a station "in person" you're usually aware of how much RF power the transceiver is generating and the Standing Wave Ratio (SWR) present at the antenna system input. This information may be provided by the transceiver if it has built-in RF power and SWR metering, or you may have a separate RF power/SWR meter nearby.

But when you're a client operator hundreds or thousands of miles away, keeping tabs on this vital data could be problematic. If you're unaware of the transceiver output level or the antenna system SWR, you are effectively operating "blind." A problem could arise in the antenna system, for instance, and the SWR could go sky high. The transceiver will respond by reducing its output power to almost nothing. If you are oblivious to this fact, you'll call station after station and wonder why no one seems to hear you. Worse yet, long term operation into a high SWR could potentially damage the radio.

The obvious solution is to share this data with the client. Fortunately, modern transceivers with computer control capability often make output power and SWR metering data available to the outside world. If your chosen rig control software can display this information, the client will be able to see it as well. Keep this functionality in mind as you shop for host/client or transceiver control software. Some programs will only display SWR and RF power data when used with certain radios. And if you'll be investing in a new transceiver for your host station, make sure it is fully compatible with your desired software. That is to say, the software should be able to display all the information the transceiver provides.

The alternative is to purchase a separate SWR/Power meter that can be connected to the host computer. Several of these computer-capable meters exist. One example is the Wavenode (**www.wavenode.com**), which includes remote monitoring software to give the client operator a highly detailed sense of how the host station is performing.

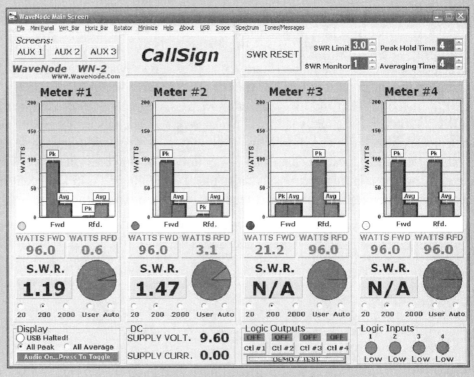

Wavenode remote monitoring software.

Appendix

A Trio of Transceiver/Computer Interfaces

Virtually all modern Amateur Radio transceivers (and many general-coverage receivers) have provisions for external computer control. Most hams take advantage of this feature using software specifically developed for control, or primarily intended for some other purpose (such as contest logging), with rig control as a secondary function.

Unfortunately, the serial port on most radios cannot be directly connected to the serial port on most computers. The problem is that most radios use TTL signal levels while most computers use RS-232-D.

The interfaces described here simply convert the TTL levels used by the radio to the RS-232-D levels used by the computer, and vice versa. Interfaces of this type are often referred to as level shifters. Two basic designs, one having a couple of variations, cover the popular brands of radios. This article, by Wally Blackburn, AA8DX, first appeared in February 1993 *QST*.

TYPE ONE: ICOM CI-V

The simplest interface is the one used for the ICOM CI-V system. This interface works with newer ICOM and Ten-Tec rigs. **Figure 1** shows the two-wire bus system used in these radios.

This arrangement uses a CSMA/CD (carrier-sense multiple access/collision detect) bus. This refers to a bus that a number of stations share to transmit and receive data. In effect, the bus is a single wire and common ground that interconnect a number of radios and computers.

The single wire is used for transmitting and receiving data. Each device has its own unique digital address. Information is transferred on the bus in the form of packets that include the data and the address of the intended receiving device.

The schematic for the ICOM/Ten-Tec interface is shown in **Figure 2**. It is also the Yaesu interface. The only difference is that the transmit data (TxD) and receive data (RxD) are jumpered together for the ICOM/Ten-Tec version.

The signal lines are active-high TTL. This means that a logical one is represented by a binary one (+5 V). To shift this to RS-232-D it must converted to –12 V while a binary zero (0 V) must be converted to +12 V. In the other direction, the opposites are needed: –12 V to +5 V and +12 V to 0 V.

U1 is used as a buffer to meet the interface specifications of the radio's circuitry and provide some isolation. U2 is a 5-V-powered RS-232-D transceiver chip that translates between TTL and RS-232-D levels. This chip uses charge pumps to obtain ±10 V from a single +5-V supply. This device is used in all three interfaces.

A DB25 female (DB25F) is typically used at the computer end. Refer to the discussion of RS-232-D earlier in the chapter for 9-pin connector information. The interface connects to the radio via a 1/8-inch phone plug. The sleeve is ground and the tip is the bus connection.

It is worth noting that the ICOM and Ten-Tec radios use identical basic command sets (although the Ten-Tec includes additional commands). Thus, driver software is compatible. The manufacturers are to be commended for working toward standardizing these interfaces somewhat. This allows Ten-Tec radios to be used with all

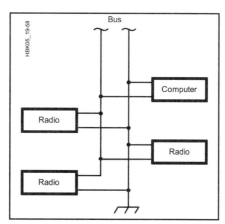

Figure 1 — The basic two-wire bus system that ICOM and newer Ten-Tec radios share among several radios and computers. In its simplest form, the bus would include only one radio and one computer.

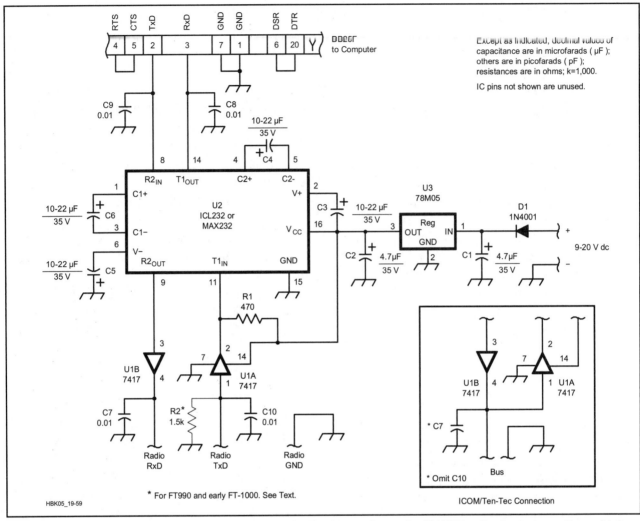

Figure 2 — ICOM/Ten-Tec/Yaesu interface schematic. The insert shows the ICOM/Ten-Tec bus connection, which simply involves tying two pins together and eliminating a bypass capacitor.

C7-C10 — 0.01-µF ceramic disc.   U1 — 7417 hex buffer/driver.   U2 — Harris ICL232 or Maxim MAX232.

popular software that supports the ICOM CI-V interface. When configuring the software, simply indicate that an ICOM radio (such as the IC-735) is connected.

TYPE TWO: YAESU INTERFACE

The interface used for Yaesu rigs is identical to the one described for the ICOM/Ten-Tec, except that RxD and TxD are not jumpered together. Refer to Figure 2. This arrangement uses only the RxD and TxD lines; no flow control is used.

The FT-990 and FT-1000/FT-1000D use an emitter follower as the TTL output. If the input impedance of the interface is not low enough to pull the input below threshold with nothing con- nected, the FT-990/FT-1000 (and perhaps other radios) will not work reliably. This can be corrected by installing a 1.5 kΩ resistor from the serial out line to ground (R2 on pin 1 of U1A in Figure 2.) Later FT-1000D units have a 1.5 kΩ pulldown resistor incorporated in the radio from the TxD line to ground and may work reliably without R2.

The same computer connector is used, but the radio connector varies with model. Refer to the manual for your particular rig to determine the connector type and pin arrangement.

TYPE THREE: KENWOOD

The interface setup used with Kenwood radios

2 Appendix

is different in two ways from the previous two: Request-to-Send (RTS) and Clear-to-Send (CTS) handshaking is implemented and the polarity is reversed on the data lines. The signals used on the Kenwood system are active-low. This means that 0 V represents a logic one and +5 V represents a logic zero. This characteristic makes it easy to fully isolate the radio and the computer since a signal line only has to be grounded to assert it. Opto-isolators can be used to simply switch the line to ground.

The schematic in **Figure 3** shows the Kenwood interface circuit. Note the different grounds for the computer and the radio. This, in conjunction with a separate power supply for the interface, provides excellent isolation.

The radio connector is a 6-pin DIN plug. The manual for the rig details this connector and the pin assignments.

Some of the earlier Kenwood radios require additional parts before their serial connection can be used. The TS-440S and R-5000 require installation of a chipset and some others, such as the TS-940S require an internal circuit board.

CONSTRUCTION AND TESTING

The interfaces can be built using a PC board,

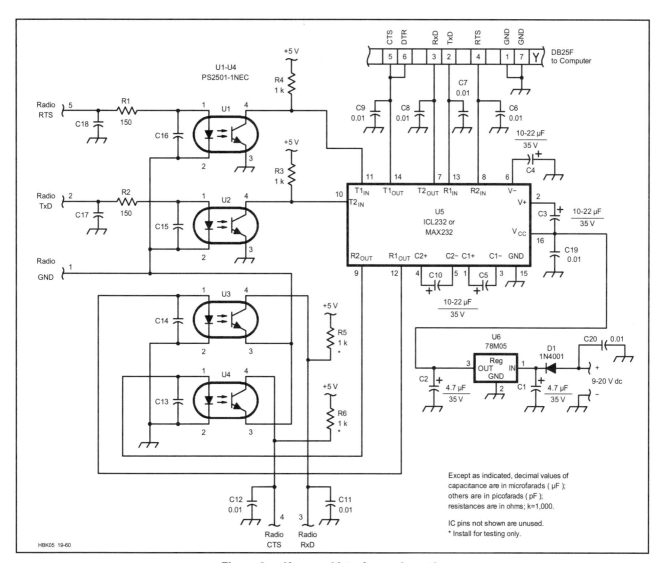

Figure 3 — Kenwood interface schematic.

C6-C9, C11, C12, C17, C18 — 0.01-µF ceramic disc.
C13-C16, C19-C21 — 0.01 µF ceramic disc.

U1-U4 — PS2501-1NEC (available from Digi-Key).

U5 — Harris ICL232 or Maxim MAX232.

Table 1
Kenwood Interface Testing

Apply	Result
GND to Radio-5	−8 to −12 V at PC-5
+5 V to Radio-5	+8 to +12 V at PC-5
+9 V to PC-4	+5 V at Radio-4
−9 V to PC-4	0 V at Radio-4
GND to Radio-2	−8 to −12 at PC-3
+5 V to Radio-2	+8 to +12 V at PC-3
+9 V to PC-2	+5 V to Radio-3
−9 V to PC-2	0 V at Radio-3

Table 2
ICOM/Ten-Tec Interface Testing

Apply	Result
GND to Bus	+8 to +12 V at PC-3
+5 V to Bus	−8 to −12 V at PC-3
−9 V to PC-2	+5 V on Bus
+9 V to PC-2	0 V on Bus

Table 3
Yaesu Interface Testing

Apply	Result
GND to Radio TxD	+8 to +12 V at PC-3
+5 V to Radio TxD	−8 to −12 V at PC-3
+9 V to PC-2	0 V at Radio RxD
−9 V to PC-2	+5 V at Radio RxD

breadboarding, or point-to-point wiring. PC boards and MAX232 ICs are available from FAR Circuits. Use the *TIS Find* program for the latest address information. The PC board template is available on the CD-ROM.

It is a good idea to enclose the interface in a metal case and ground it well. Use of a separate power supply is also a good idea. You may be tempted to take 13.8 V from your radio — and it works well in many cases: but you sacrifice some isolation and may have noise problems. Since these interfaces draw only 10 to 20 mA, a wall transformer is an easy option.

The interface can be tested using the data in **Tables 1, 2** and **3**. Remember, all you are doing is shifting voltage levels. You will need a 5-V supply, a 9-V battery and a voltmeter. Simply supply the voltages as described in the corresponding table for your interface and check for the correct voltage on the other side. When an input of −9 V is called for, simply connect the positive terminal of the battery to ground.

During normal operation, the input signals to the radio float to 5 V because of pullup resistors inside the radio. These include RxD on the Yaesu interface, the bus on the ICOM/Ten-Tec version, and RxD and CTS on the Kenwood interface. To simulate this during testing, these lines must be tied to a 5-V supply through 1-kΩ resistors. Connecting these to the supply without current-limiting resistors will damage the interface circuitry. R5 and R6 in the Kenwood schematic illustrate this. They are not shown (but are still needed) in the ICOM/Ten-Tec/Yaesu schematic. Also, be sure to note the separate grounds on the Kenwood interface during testing.

Another subject worth discussing is the radio's communication configuration. The serial ports of both the radio and the computer must be set to the same baud rate, parity, and number of start and stop bits. Check your radio's documentation and configure your software or use the PC-DOS/MS-DOS MODE command as described in the computer manual.

A Simple Serial Interface

A number of possibilities exist for building an RS-232-to-TTL interface level converter to connect a transceiver to a computer serial port. The interface described here by Dale Botkin, NØXAS, is extremely simple, requiring only a few common parts.

Most published serial interface circuits fall into one of the following categories:

• Single chip, MAX232 type. These are relatively expensive, require several capacitors, take up lots of board space and usually require ordering parts.

• Single chip, MAX233 type. This version doesn't need capacitors. Otherwise it shares most of the characteristics of the MAX232 type device, but is even more expensive.

• Transistors. Usually transistor designs use a mix of NPN and PNP devices, four to eight or more resistors, and usually capacitors and diodes. This is just too many parts!

• Other methods. Some designs use series resistors, or TTL logic inverters with resistors. The idea is that the voltages may be out of specification for the chip, but if the current is low enough it won't damage the device. This is not always safe for the TTL/CMOS device. With just a series resistor the serial data doesn't get inverted, so it requires the use of a software USART. It can work, but it's far from ideal.

A SIMPLER APPROACH

The EIA-232 specification specifies a signal level between ±3 V and ±12 V for data. The author looked at data sheets for devices ranging from the decades-old 1489 line receiver to the latest single-chip solutions. Not a single device found actually required a negative input voltage! In fact, the equivalent schematic shown for the old parts clearly shows a diode that holds the input to the first inverter stage at or near ground when a negative voltage is applied. The data sheets show that a reasonable TTL-level input will drive these devices perfectly well.

Relieved of the burden of exactly replicating the PC-side serial port voltages, the interface shown in **Figure 4** is designed for minimum parts count and low cost — under $1.

Why use MOSFETs for Q1 and Q2? In switching application such as this, there are several advantages to using small signal MOSFETs over the usual bipolar choices like the 2N2222 or 2N3904. First, they need no base current limiting resistors — the MOSFET is a voltage operated device, and the gate is isolated from the source and drainch The gate on a 2N7000 is rated for ±20 V, so you don't need a clamping diode to prevent negative voltages from reaching the gate. They can also be had just as cheaply as their common bipolar cousins. (If sourcing 2N7000s is a challenge in Europe, the BS170 is basically identical except for the pins being reversed.) Now all that is needed is a drain limiting resistor, since the gate current is as close to zero as we are likely to care about.

Q1 gets its input from the TTL-level device. Q1 and R1 form a simple inverter circuit, with the inverted output going to pin 2 (RxD) of the serial connector. The pull-up voltage for RxD comes from pin 7 (RTS) of the serial connector, so the

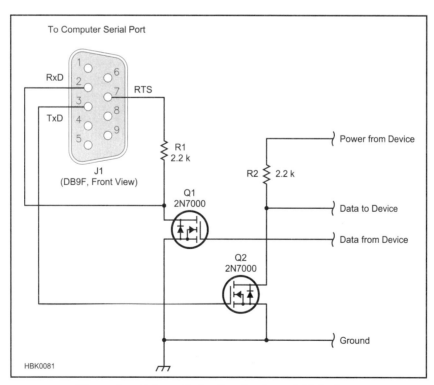

Figure 4 — Schematic for the simple serial interface.

J1 — DB9F or connector to fit device serial port. Note that pin assignments for RxD, TxD and RTS are different on DB25 connectors.
Q1, Q2 — Small signal MOSFET, 2N7000 or equiv.

voltage will be correct for whatever device is connected — be it a PC or whatever else you have. Q2 and R2 are another inverter, driven by TxD from pin 3 of the serial connector. The pull-up voltage for R2 should be supplied by the low-voltage device — your radio or microcontroller project.

There you go. Two cheap little MOSFETs, two resistors. The 2.2 kΩ, ¼ W resistors, seem to work with every serial port tried so far. The author has sold several hundred kits using this exact circuit for the serial interface without issues from any users. It's been tested with laptops, desktops, USB-to serial-converters, a serial LCD display and even a Palm-III with an RS-232 serial cable. It's also been used with very slight modifications in computer interfaces for VX-7R and FT-817 transceivers. It's simple, cheap and works great — what more could you ask?

USB Interfaces For Your Ham Gear

Over the past several years, serial ports have been slowly disappearing from computers. The once-ubiquitous serial port, along with other interfaces that were once a basic part of every computer system, has largely been replaced by Universal Serial Bus (USB) ports. This presents problems to the amateur community because serial ports have been and continue to be the means by which we connect radios to computers for control, programming, backing up stored memory settings, contest logging and other applications. Radios equipped with USB ports are rare indeed, and many of the optional computer interface products sold by ham manufacturers are serial-only. The project described here by Dale Botkin, NØXAS, offers a solution.

THE UNIVERSAL SERIAL BUS

So what is USB, and how can it be used in the ham shack? The USB standard came about during the early 1990s as a replacement for the often inconvenient and frustrating array of serial, parallel, keyboard, mouse and various other interfaces commonly used in personal computer systems. The idea is to use a single interface that can handle multiple devices sending and receiving different types of data, at different rates, over a common bus. In practice, it works quite well because of the very well defined industry specification for signal levels, data formats, software drivers and so forth.

At least two and sometimes as many as six or more USB ports are now found as standard equipment on nearly all new computer systems. Support for USB ports and generic USB devices is built into most operating systems including *Windows*, *Linux*, *Solaris*, *Mac OS* and *BSD* to name a few. USB use has spread beyond computers, and the standard USB "B" or the tiny 4-conductor "Mini B" connectors are now found on all manner of electronic devices including cameras, cell phones, PDAs, TVs, CD and DVD recorders, external data storage drives and even digital picture frames. USB is indeed becoming *universal*, as its name implies — everywhere, it seems, except for Amateur Radio gear!

The EIA-232 serial interface specification we have used for decades is just that — an electrical interface specification that defines voltage levels used for signaling. The asynchronous serial format most commonly used with serial ports defines how data is sent over whatever interface is being used. Any two devices using the same protocol and format can talk to each other, whether they are both computers, both peripheral devices (like printers, terminals or TNCs) or any combination. All that is required is the proper cable and the selection of matching parameters (speed, data bits, handshaking) on each end of the connection.

This is not the case with USB devices. USB uses a tiered-star architecture, with well defined roles for the host controller (your computer) and devices. Devices cannot communicate directly with each other, only with a host.

See **Figure 5**. At the top of the stack is the *host controller device* (HCD), usually your computer, which controls all communication within the USB network. The computer may have several host controllers, each with one or more ports. The host controller discovers, identifies and talks to up to 127 devices using a complex protocol. No device can communicate directly with another device; all communication is initiated and controlled by the

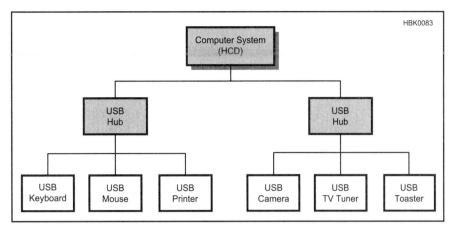

Figure 5 — Typical USB architecture.

host controller. As a result, you cannot simply connect two USB devices together and expect them to communicate. (Some USB printers will talk directly to digital cameras because the printer also has a host controller built in.)

As you have probably already figured out, moving from serial to USB is not a simple matter of converting voltage levels or even data formats. USB devices all need some embedded "smarts," usually in the form of either a dedicated USB interface chip or a microcontroller using firmware to talk to the USB host. Where we could once use simple transistor circuits to adapt RS-232 serial to the TTL voltage levels used by many radios, we now must add the complete USB interface into the picture.

The USB specification is freely available and well documented, and there are books dealing with designing the software to use the interface. If you're starting from scratch it is a very daunting task to even understand exactly how USB works, let alone design and build your own USB-equipped device. If you're willing and dedicated enough, you can implement your own USB device using a fairly broad range of available microcontrollers. There are also dedicated function chips that implement a complete USB interface — or parts of it — in one or two devices.

Fortunately, several manufacturers have stepped up to make using USB relatively easy for the experimenter. One of the most commonly used solutions is from Future Technology Devices International (FTDI). FTDI manufactures a line of USB interface chips that plug into the Universal Serial Bus on one side and present a serial or parallel interface on the other side. The chips are quite versatile and are commonly used to add USB functionality to a wide range of devices ranging from cell phones to USB-to-serial cables. They are also available in several forms such as plug-in modules with a USB connector and the interface logic, all on a small PC board with pins that can be plugged into a standard IC socket.

CONSTRUCTION

This project uses an FT232R series chip from FTDI to bridge the USB interface from your computer to a TTL-level serial interface that can be adapted to use with many different ham rigs. The chip has built-in logic to handle the USB interface, respond to queries and commands from the computer, and send and receive serial data. By using special drivers available from FTDI (and built into *Windows* and many other systems), it can appear to your programs as a standard serial COM port.

Once the conversion to serial data is done, all that remains is to adapt the signals to the particular interface used by your equipment. In many cases, this serial data can simply be connected directly to the rig using the appropriate connector. In others it may require some extra circuitry to accommodate various interface schemes used by manufacturers. The most common is an open-collector bus type interface, which is easily accommodated using a few inexpensive parts.

Construction of this project can be handled in a couple of different ways. The FT232R series chips are available from various suppliers at low cost. Unfortunately they are only available in SSOP and QFLP surface mount packages that can be difficult to work with unless you use a custom printed circuit board. (An additional letter is tacked onto the part number to identify the package. For example, the FT232RL uses a 28-pin SSOP package.) There are SSOP-to-through-hole adapter boards available, but they are relatively expensive and still require some fine soldering.

Fortunately FTDI also sells a solution in the form of the TTL-232R (**Figure 6**). It incorporates a USB "A" connector, an FT232RQ chip and the required power conditioning on a tiny PCB encapsulated in plastic, all about the size of a normal USB connector. A 6-ft long cable terminates in a 6-position inline header that provides ground, +5 V, and TTL level TxD, RxD, RTS and CTS signals.

You can connect these signals directly to your own microprocessor projects and have an "instant" USB connection. You can also build a variety of interface cables or boards for use with different radios. This arrangement is ideal for the experimenter and the person who might need several different interfaces for use with several different rigs — like many of us! The cost of the TTL-232R cable is quite reasonable, in fact lower than the cost of having a prototype PCB made to build your own equivalent device. Of course that option is still available to those who need more flexibility.

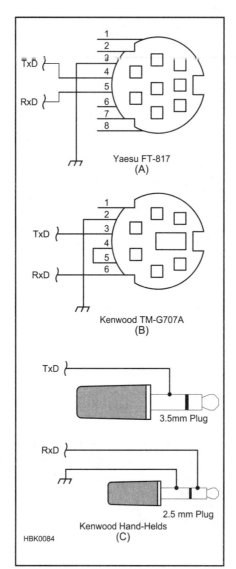

Figure 7 — Using the FTDI USB-to-serial interface with some radios is as simple as wiring up a connector. Connections for the Yaesu FT-817 are shown at A, the Kenwood TM-G707A (PG-4 type) at B, and Kenwood hand-helds (PG-5 type) at C. Ground, RxD and TxD connections are made to the 6-pin connector on the TTL-232R cable.

Figure 6 — The FTDI TTL-232R cable simplifies construction.

USING USB IN THE SHACK

With the USB-to-serial conversion handled so easily, the only remaining task is to adapt the TTL-level serial interface of the FT232R to the interface used by your particular radio. In the case of some rigs, such as the Yaesu FT-817, Kenwood TM-G707A and many others, it's simply a matter of wiring up the right connector, as shown in **Figure 7**.

For communicating with individual ICOM and Ten-Tec rigs, one simple solution is to simply connect the transmit and receive data signals together and wire up a ⅛-inch stereo plug as shown in **Figure 8**. This will work with a single device connected, but lacks the open-collector drivers needed to be compatible with attaching multiple devices to the CI-V bus.

A LITTLE MORE VERSATILITY

Many hand-held transceivers use a single data line for both transmit and receive. This method is very similar to CI-V, but requires an open-collector or open-drain configuration for a single device. Some additional buffering is required to provide an interface to the rig.

Figure 9 shows the design of the general-purpose buffer board for use with various radios. This circuit simply provides a noninverting, open-drain buffer for data in each direction. Adapters for various rigs can be plugged into J2 with no further changes needed. Use of MOSFETs instead of their bipolar counterparts simply reduces the parts count. No additional resistors are needed to limit base current, as you would need when using 2N2222 or similar transistors. Parts can be mounted on a small project board as shown in **Figure 10**.

This arrangement turns out to be extremely versatile and will work with CI-V, FT-817, Yaesu VX-series handhelds and the other rigs the author was able to test as well. All that is required is the proper adapter cable from the base interface board to the rig. Some examples are shown in **Figure 11**.

ADDING PTT

Handling PTT is quite simple, and it can be done either with or without isolation as shown in **Figure 12** (on Appendix page 11). For a non-isolated PTT line, we can simply drive the gate of a 2N7000 or similar MOSFET from the FT232RL's RTS output. If you prefer an

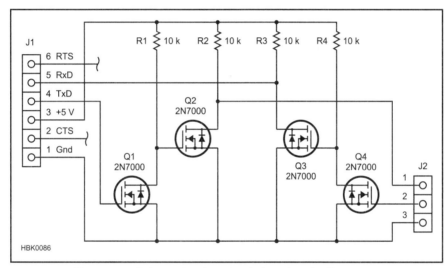

Figure 8 — The transmit and receive data signals can be connected together for use with a single Ten-Tec or ICOM CI-V device.

Figure 9 — Schematic of the basic interface buffer board.
J1 — 6-pin, 0.1-in pitch header to match TTL-232R cable.
J2 — 3-pin header to match radio cable.
Q1-Q4 — Small-signal MOSFET, 2N7000, BS-170 or equiv.
R1-R4 — 10 kΩ, ¼ W.

Figure 10 — The completed buffer board. Adapter cables for various radios plug into the header on the right edge of the board.

optically isolated PTT output, all you need to add is one resistor and an optocoupler such as a 4N27 or PS-7142. Note that the TTL-232R provides a 100 Ω resistor in series with the RTS signal (the same is true for TxD), so a smaller value resistor (R1) can be used for the optocoupler's input than you would normally use.

While connecting the data lines directly to the rig is a suitable solution for many applications, some Kenwood HF rigs require a different approach. They require inverted TTL-level data, and may experience "hash" noise from the computer when not using an isolated interface. **Figure 13** (next page) shows an optically isolated method of connecting to a Kenwood rig such as the TS-850S. Note that this has not been tested but the design is similar to numerous examples of serial interfaces for the TS-850S and similar radios.

It's probably only a matter of time before we start seeing the tiny USB "Mini B" connector appear on new ham gear. It has become inexpensive and easy to add USB to pretty much any product, and as serial ports become more and more rare the manufacturers will eventually face overwhelming consumer demand to switch. In the meantime, you can do a little homebrewing and use your new, serial port free computer to talk to all of your ham gear for very little cost.

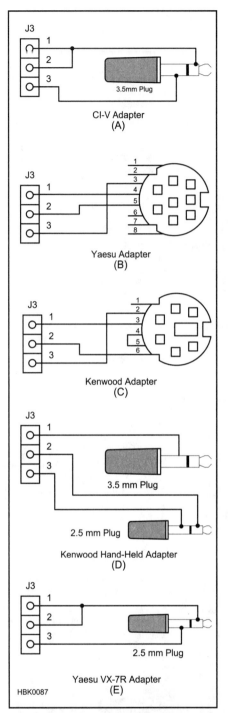

Figure 11 — Adapters for using the buffer board with various radios. J3 matches the adapter used on the buffer board.

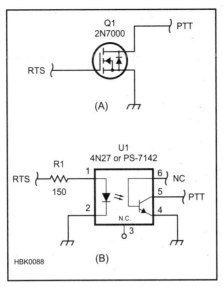

Figure 12 — PTT can be added either without isolation (A) or with an optoisolator (B).

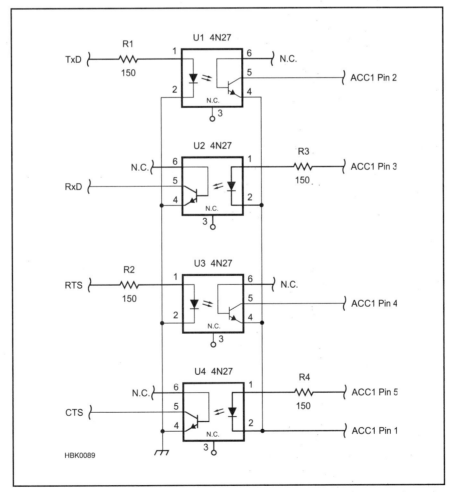

Figure 13 — Isolated Kenwood interface for the TS-850S and similar HF transceivers requiring inverted TTL-level data and isolated lines.

Appendix 11

Index

A
Address
   Dynamic: 2-10
   IP: 2-8
   Network: 2-7
   Static: 2-10
Amplifier: 4-13
Antenna
   Multiband: 7-12
   Rotator Control: 4-11, 7-14
   Switching: 4-11, 7-13
ARTiE (software): 7-7
Audio
   Computer Connections: 4-2
   EchoLink: 5-6
   Levels: 4-5, 5-5
   Quality: 5-5
   Receive: 4-4
   Skype: 5-3
   Sound Devices: 4-3
   Streaming
      Public: 6-2
         Single Stream: 6-7
   Telephone: 5-6
   Transmit: 4-5
   VoIP: 5-2
Awards and Contests by Remote Control: 1-8

C
CAT (Computer Aided Transceiver): 4-10
Client Hardware: 4-16
CW by Remote Control: 4-16, 7-15
   CWType: 7-15
   WinKeyer: 7-16

D
DHCP: 3-11
Digital Modes by Remote Control: 7-10
DNS: 2-9
DSL: 2-3
DXCC and Remote Control: 1-8
DXLab Commander (software): 7-6

E
EchoLink: 5-6, 7-7
Ethernet: 4-10
   Adapters for Remote Control: 7-17, 7-19

F
Failsafe Shutdown: 4-13
FCC
   And Internet Remote Control: 1-6
Fiber Optic Internet: 2-4
Firewalls: 3-5
   Opening Ports: 3-9

G
GlenTek Ethernet Adapter for Remote Control: 7-17
GoToMyPC: 7-10

H
Ham Radio Deluxe (software): 7-7
Hamachi VPN (software): 7-4

I
Interfaces
   Sound Card: 4-3
   Transceiver Control: 4-10
      Build Your Own: .... Appendix
Internet
   Addresses: 2-7
   Cable TV: 2-3
   Dial-Up: 2-2
   DSL: 2-3
   Fiber Optic: 2-4
   Satellite: 2-4
   Speed: 2-2
   Wireless: 2-5
Internet Remote Base (software): 7-5
IP Address: 2-8

L
LAN: 2-7
LogMeIn: 7-10

M
Microphone/Headset: 4-16
Mocha (software): 7-10
Monitoring SWR and Power Remotely: 7-20

N

Networks: 2-6
    Addresses: 2-8
        Dynamic: 2-10
        IP: 2-8
        Static: 2-10

P

PCAnywhere (software): 7-10
Pings: 3-6
Ports
    Addressing: 3-2
        TCP: 3-2
        UDP: 3-2
    COM Port Redirection Software: 7-4
    Ethernet: 4-10
    Forwarding: 3-10
    Opening: 3-9
    RS-232: 4-10
    USB: 4-10
    Virtual Serial: 4-9
Power Output Monitoring: 7-20
PTT: 4-7

R

RealVNC: 7-10
Receiver, Remote: 6-2
    Tunable: 6-7
    Web: 6-9
RemoteRig Ethernet Adapter: 7-19
Rotator Control: 4-11, 7-14
Routers: 3-5, 3-8
    DHCP: 3-11
    Port Forwarding: 3-10
RS-232: 4-10
Rules, FCC: 1-6

S

Satellite Internet: 2-4
Security: 3-5
    Anti-Virus: 3-6
    Failsafe Shutdown: 4-13
    Firewalls: 3-6
    Routers: 3-9
Software
    ARTiE: 7-7
    COM Port Redirectors: 7-4
    DXLab Commander: 7-6
    EchoLink: 5-6, 7-7
    Ham Radio Deluxe: 7-7
    Hamachi Virtual Private Network: 7-4
    Internet Remote Base: 7-5
    Internet Remote Toolkit: 7-3
    MixW: 7-3
    Mocha: 7-10
    PCAnywhere: 7-10
    RealVNC: 7-10
    TightVNC: 7-10
    TRX Manager: 7-6
    VoIP: 5-2
Sound Cards/Devices: 4-3
    External: 4-7
Station Configurations: 7-11
    Power and SWR Monitoring: 7-20
Streaming Audio: 6-2
Switching
    Antennas: 4-11
    Transmit/Receive: 4-7
SWR, Monitoring Remotely: 7-20

T

TCP: 3-2
Telnet: 7-6
TightVNC (software): 7-10
Transceiver
    Frequency Control: 4-10
    Transmit/Receive Switching: 4-7
Transmit Audio: 4-5
Transmit/Receive Switching: 4-7
TRX Manager (software): 7-6
TTL: 4-10

U

UDP: 3-2
USB: 4-10

V

VHF/UHF Century Club and Remote Control: 1-8
Video: 7-17
Virtual Ports: 4-9
Virus, Computer: 3-6
VNC (Virtual Network Computing): 7-8
    GoToMyPC: 7-10
    LogMeIn: 7-10
    Mocha: 7-10
    PCAnywhere: 7-10
    RealVNC: 7-10
    TightVNC: 7-10
VoIP: 5-2
    Audio Levels: 5-5
    Call Placement and Answering: 5-6
    EchoLink: 5-6, 7-7
    Skype: 5-3
VPN (Virtual Private Network): 7-4

W

WiFi: 2-5
WiMAX: 2-5
Windows
    Audio Control: 4-5
WinKeyer: 7-16
Worked All States and Remote Control: 1-8

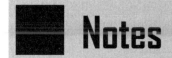

Notes

About the ARRL

The seed for Amateur Radio was planted in the 1890s, when Guglielmo Marconi began his experiments in wireless telegraphy. Soon he was joined by dozens, then hundreds, of others who were enthusiastic about sending and receiving messages through the air—some with a commercial interest, but others solely out of a love for this new communications medium. The United States government began licensing Amateur Radio operators in 1912.

By 1914, there were thousands of Amateur Radio operators—hams—in the United States. Hiram Percy Maxim, a leading Hartford, Connecticut inventor and industrialist, saw the need for an organization to band together this fledgling group of radio experimenters. In May 1914 he founded the American Radio Relay League (ARRL) to meet that need.

Today ARRL, with approximately 155,000 members, is the largest organization of radio amateurs in the United States. The ARRL is a not-for-profit organization that:
- ✔ promotes interest in Amateur Radio communications and experimentation
- ✔ represents US radio amateurs in legislative matters, and
- ✔ maintains fraternalism and a high standard of conduct among Amateur Radio operators.

At ARRL headquarters in the Hartford suburb of Newington, the staff helps serve the needs of members. ARRL is also International Secretariat for the International Amateur Radio Union, which is made up of similar societies in 150 countries around the world.

ARRL publishes the monthly journal *QST*, as well as newsletters and many publications covering all aspects of Amateur Radio. Its headquarters station, W1AW, transmits bulletins of interest to radio amateurs and Morse code practice sessions. The ARRL also coordinates an extensive field organization, which includes volunteers who provide technical information and other support services for radio amateurs as well as communications for public-service activities. In addition, ARRL represents US amateurs with the Federal Communications Commission and other government agencies in the US and abroad.

Membership in ARRL means much more than receiving *QST* each month. In addition to the services already described, ARRL offers membership services on a personal level, such as the Technical Information Service—where members can get answers by phone, email or the ARRL website, to all their technical and operating questions.

Full ARRL membership (available only to licensed radio amateurs) gives you a voice in how the affairs of the organization are governed. ARRL policy is set by a Board of Directors (one from each of 15 Divisions). Each year, one-third of the ARRL Board of Directors stands for election by the full members they represent. The day-to-day operation of ARRL HQ is managed by an Executive Vice President and his staff.

No matter what aspect of Amateur Radio attracts you, ARRL membership is relevant and important. There would be no Amateur Radio as we know it today were it not for the ARRL. We would be happy to welcome you as a member! (An Amateur Radio license is not required for Associate Membership.) For more information about ARRL and answers to any questions you may have about Amateur Radio, write or call:

ARRL — the national association for Amateur Radio
225 Main Street
Newington CT 06111-1494
Voice: 860-594-0200
   Fax: 860-594-0259
   E-mail: **hq@arrl.org**
   Internet: **www.arrl.org/**

Prospective new amateurs call (toll-free):
800-32-NEW HAM (800-326-3942)
You can also contact us via e-mail at **newham@arrl.org**
or check out *ARRLWeb* at **www.arrl.org/**

FEEDBACK

Please use this form to give us your comments on this book and what you'd like to see in future editions, or e-mail us at **pubsfdbk@arrl.org** (publications feedback). If you use e-mail, please include your name, call, e-mail address and the book title, edition and printing in the body of your message. Also indicate whether or not you are an ARRL member.

Where did you purchase this book?
  ☐ From ARRL directly   ☐ From an ARRL dealer

Is there a dealer who carries ARRL publications within:
  ☐ 5 miles   ☐ 15 miles   ☐ 30 miles   of your location?   ☐ Not sure.

License class:
  ☐ Novice   ☐ Technician   ☐ Technician with code   ☐ General   ☐ Advanced   ☐ Amateur Extra

Name _____ ARRL member? ☐ Yes ☐ No
_____ Call Sign _____

Daytime Phone (___) _____ Age _____

Address _____

City, State/Province, ZIP/Postal Code _____

If licensed, how long? _____ e-mail address: _____

Other hobbies _____

_____

Occupation _____

For ARRL use only	Remote
Edition	1 2 3 4 5 6 7 8 9 10 11 12
Printing	1 2 3 4 5 6 7 8 9 10 11 12